THE CIRCULAR ECONOMY AND THE GLOBAL SOUTH

The circular economy is a policy approach and business strategy that aims to improve resource productivity, promote sustainable consumption and production, and reduce environmental impacts. This book examines the relevance of the circular economy in the context of developing countries, something which to date is little understood.

This volume highlights examples of circular economy practices in developing country contexts in relation to small and medium enterprises (SMEs), informal sector recycling and national policy approaches. It examines a broad range of case studies, including Argentina, Brazil, China, Colombia, India, Indonesia, Kenya, South Africa and Thailand, and illustrates how the circular economy can be used as a new lens and possible solution to cross-cutting development issues of pollution and waste, employment, health, urbanisation and green industrialisation. In addition to more technical and policy-oriented contributions, the book also critically discusses existing narratives and pathways of the circular economy in the Global North and South, and how these differ or possibly even conflict with each other. Finally, the book critically examines under what conditions the circular economy will be able to reduce global inequalities and promote human development in the context of the Sustainable Development Goals.

Presenting a unique social sciences perspective on the circular economy discourse, this book is relevant to students and scholars studying sustainability in economics, business studies, environmental politics and development studies.

Patrick Schröder is a Research Fellow at the Institute of Development Studies, UK and holds a PhD from Victoria University of Wellington, New Zealand.

Manisha Anantharaman is an Assistant Professor in Justice, Community and Leadership at Saint Mary's College of California, USA and holds a PhD from University of California Berkeley, USA.

Kartika Anggraeni is a project manager at the Collaborating Centre for Sustainable Consumption and Production (CSCP) in Wuppertal, Germany. She is currently co-managing an EU SWITCH Africa Green project in Kenya.

Timothy J. Foxon is Professor of Sustainability Transitions at the University of Sussex, UK and the author of *Energy and Economic Growth: why we need a new pathway to prosperity* (Routledge, 2017).

PATHWAYS TO SUSTAINABILITY

This book series addresses core challenges around linking science and technology and environmental sustainability with poverty reduction and social justice. It is based on the work of the Social, Technological and Environmental Pathways to Sustainability (STEPS) Centre, a major investment of the UK Economic and Social Research Council (ESRC). The STEPS Centre brings together researchers at the Institute of Development Studies (IDS) and SPRU (Science Policy Research Unit) at the University of Sussex with a set of partner institutions in Africa, Asia and Latin America.

Series Editors:
Ian Scoones and Andy Stirling
STEPS Centre at the University of Sussex

Editorial Advisory Board:
Steve Bass, Wiebe E. Bijker, Victor Galaz, Wenzel Geissler, Katherine Homewood, Sheila Jasanoff, Melissa Leach, Colin McInnes, Suman Sahai, Andrew Scott

Titles in this series include:

One Health
Science, politics and zoonotic disease in Africa
Edited by Kevin Bardosh

Grassroots Innovation Movements
Adrian Smith, Mariano Fressoli, Dinesh Abrol, Elisa Around and Adrian Ely

Agronomy for Development
The Politics of Knowledge in Agricultural Research
James Sumberg

The Water-Food-Energy Nexus
Power, Politics and Justice
Jeremy Allouche, Carl Middleton and Dipak Gyawali

The Circular Economy and the Global South
Sustainable Lifestyles and Green Industrial Development
Edited by Patrick Schröder, Manisha Anantharaman, Kartika Anggraeni and Tim Foxon

THE CIRCULAR ECONOMY AND THE GLOBAL SOUTH

Sustainable Lifestyles and Green Industrial Development

Edited by Patrick Schröder, Manisha Anantharaman, Kartika Anggraeni and Timothy J. Foxon

First published 2019
by Routledge
2 Park Square, Milton Park, Abingdon, Oxon OX14 4RN

and by Routledge
52 Vanderbilt Avenue, New York, NY 10017

Routledge is an imprint of the Taylor & Francis Group, an informa business

© 2019 selection and editorial matter, Patrick Schröder, Manisha Anantharaman, Kartika Anggraeni and Timothy J. Foxon; individual chapters, the contributors

The right of Patrick Schröder, Manisha Anantharaman, Kartika Anggraeni and Timothy J. Foxon to be identified as the authors of the editorial material, and of the authors for their individual chapters, has been asserted in accordance with sections 77 and 78 of the Copyright, Designs and Patents Act 1988.

All rights reserved. No part of this book may be reprinted or reproduced or utilised in any form or by any electronic, mechanical, or other means, now known or hereafter invented, including photocopying and recording, or in any information storage or retrieval system, without permission in writing from the publishers.

Trademark notice: Product or corporate names may be trademarks or registered trademarks, and are used only for identification and explanation without intent to infringe.

British Library Cataloguing in Publication Data
A catalogue record for this book is available from the British Library

Library of Congress Cataloging-in-Publication Data
A catalog record has been requested for this book

ISBN: 978-1-138-35892-8 (hbk)
ISBN: 978-1-138-35893-5 (pbk)
ISBN: 978-0-429-43400-6 (ebk)

Typeset in Bembo
by Taylor & Francis Books

CONTENTS

List of illustrations viii
List of contributors x
Acknowledgements xvi
Abbreviations xviii
Preface xx

PART I
Introduction   **1**

1 Introduction: Sustainable lifestyles, livelihoods and the circular economy   3
  Patrick Schröder, Manisha Anantharaman, Kartika Anggraeni, Timothy J. Foxon and Jeffrey Barber

PART II
Narratives and politics of waste and the circular economy in the Global South   **23**

2 The many circuits of a circular economy   25
  Ashish Chaturvedi, Jai Kumar Gaurav and Pragya Gupta

3 The politics of marine plastics pollution   43
  Patrick Schröder and Victoria Chillcott

4 Circular economy and inclusion of informal waste pickers: Political economy perspectives from India and Brazil 57
Patricia Noble

5 The role of women in upcycling initiatives in Jakarta, Indonesia: A case for the circular economy in a developing country 75
Priliantina Bebasari

PART III
Policy frameworks and green industrial development approaches **93**

6 The Argentinean zero waste framework: Implementation gaps and over-sight of reusable menstrual management technologies 95
Jacqueline Gaybor and Henry Chavez

7 Assessment of the circular economy transition readiness at a national level: The Colombian case 113
Claudia Lorena Garcia and Steve Cayzer

8 Promoting industrial symbiosis in China's industrial parks as a circular economy strategy: The experience of the TEDA Eco Centre 134
An Chen, Yuyan Song and Kartika Anggraeni

9 Accelerating the transition to a circular economy in Africa: Case studies from Kenya and South Africa 152
Peter Desmond and Milcah Asamba

PART IV
Livelihoods and traditional circular economy practices **173**

10 Securing nutrition through the revival of circular lifestyles: A case study of endogenous rural communities in Rajasthan 175
Deepak Sharma and Jayesh Joshi

11 Contesting thoughts and attitudes to 'Sufficiency': Organic farming in an urbanised village in Thailand 188
Atsushi Watabe

PART V
Conclusion and outlook: circular economy approaches for the Sustainable Development Goals **201**

12 Conclusion: Pathways to an inclusive circular economy 203
    Patrick Schröder, Manisha Anantharaman, Kartika Anggraeni and Timothy J. Foxon

Index *212*

ILLUSTRATIONS

Figures

2.1	Actor-based circular economy conceptual framework.	31
3.1	Line graph demonstrating the use of the terms relating to plastic referenced in the UK parliament.	45
3.2	A bar chart demonstrating the increase in media attention around plastic pollution. Online articles counted. Note the references to the *Blue Planet II* programme and the UK government's 25-Year Environment Plan.	46
4.1	Nested systems: from recycling to Green Transformations.	58
4.2	The informal recycling system in Delhi.	61
4.3	Venn diagram highlighting the main differences and similarities of the conditions that enable IWP to close the WMC via recycling in Delhi and the SPMA.	66
7.1	Enabling framework for a CE transition.	119
8.1	Industrial symbiosis network of the automobile industry in TEDA.	138
8.2	Industrial symbiosis network for water resources in TEDA.	139
8.3	TEDA Eco Centre as waste information exchange/sharing platform in the Industrial Symbiosis project.	140
8.4	Industrial symbiosis workflow in the TBNA case.	142
8.5	Types of waste material and number of related synergies created through the IS project.	143
8.6	Whole Process Management of General Industrial Solid Waste project in TEDA.	144

8.7	Initial transfer destinations of various solid wastes generated by 25 pilot manufacturing plants in TEDA in 2012.	146
8.8	TEDA Eco Centre as clean-tech and knowledge promotion platform.	147
8.9	Eco centres as circular economy and low-carbon information exchange/sharing platforms to promote CE development in industrial parks.	148
9.1	Circular economy flow diagram for a mobile phone.	168
10.1	Linear model of agriculture.	177
10.2	Circularity aspects of the revived farming system of the endogenous community.	178
10.3	Framework and steps followed in the research programme.	181
10.4	Framework of the participatory learning and action for a circular economy approach in NSFS.	182

Tables

2.1	Actors involved in the multiple circular economy circuits.	29
2.2	Dynamics of narratives.	37
4.1	Comparison of waste recovery rates in seven cities.	59
5.1	Circular business models according to Bocken et al. (2016).	77
5.2	Mapping of CE-related initiatives identified during data collection.	80
7.1	Current state: economy, recovery of materials and innovation in Colombia.	116
7.2	CE transition readiness in Colombia.	122
8.1	ISO 14001 Environment Management System training during the IS project.	141
9.1	CE-related policies, regulations and initiatives in a selection of African countries.	158
9.2	Example CE case studies in Africa.	159
9.3	Summary of findings and circular economy solutions to mobile phones.	165

Box

9.1	Circular economy stakeholder networks in Africa.	156

CONTRIBUTORS

Manisha Anantharaman is an Assistant Professor of Justice, Community and Leadership at Saint Mary's College of California, USA. A multidisciplinary problem-driven social scientist, she studies the potential for pathways to and politics of just sustainability transitions. Her research brings together the interconnected spheres of sustainability and social justice, applying participatory and ethnographic methodologies. Her recent publications have argued that researchers need to look at power dynamics within sustainability efforts to see how these initiatives challenge or reinforce existing patterns of oppression and marginalisation, an intervention that was recognised with an Early Career Scholar Award from the Sustainable Consumption Research and Action Initiative in 2016. She is a founding member of the Global Research Forum for Sustainable Consumption and Production.

Kartika Anggraeni is a project manager at the Collaborating Centre on Sustainable Consumption and Production (CSCP) in Wuppertal, Germany. As an expert in the field of sustainable consumption and production (SCP), she is currently co-managing an EU SWITCH Africa Green project in Kenya. Previously, she worked with the EU SWITCH-Asia Network Facility from 2012–2017 to promote and mainstream SCP measures and policies across 18 Asian countries.

Milcah Asamba is a market and social researcher with Kantar Millward based in Nairobi, Kenya. She is currently pursuing a doctorate in Development Studies at Jomo Kenyatta University of Agriculture and Technology. She is also a registered Lead Environmental Expert with the National Environmental Management Authority in Kenya, a position that has enabled her to work on various environmental impact assessments and audits, which opened her mind to the threats posed by current development models and their impact on the high poverty ratios in the country. Her doctoral thesis focuses on the persistent nature of poverty in Kenya.

Jeffrey Barber is President of Integrative Strategies Forum (ISF), a non-profit organisation based in metro Washington, D.C. engaged in research and policy advocacy toward sustainable production and consumption and public participation in local and global sustainability governance. His background is in media and public opinion research, studying news and public broadcast audiences. He is currently exploring the envisioning of sustainable futures and climate change in popular culture production and consumption practices. He is a co-founder of the Global Research Forum on Sustainable Production and Consumption.

Priliantina Bebasari is an alumnus of the MA in Gender and Development at the Institute of Development Studies at the University of Sussex, UK. She is a gender and inclusion specialist and has been certified as an international gender training facilitator, Project Management for Development Professional (PMD-Pro Level 1), and is also trained in gender in emergencies. Her professional experiences include research, gender mainstreaming, and monitoring and evaluation, for development programmes and organisations. She is vegan and believes in a holistic well-being that covers love to self, others, and the planet.

Steve Cayzer is a Senior Teaching Fellow in the Department of Mechanical Engineering at the University of Bath. He manages the Innovation and Technology Management MSc and contributes to various other programmes across Engineering and Management. Steve's background is in IT, particularly web information management, artificial intelligence and sustainability, and he has worked for (among others) Logica, Hewlett-Packard Laboratories, University of Bristol and the University of the West of England.

Ashish Chaturvedi has over 15 years of professional experience and specialises in climate change mitigation, waste management, sustainable consumption and production, and environmental policy. He currently is Director Climate Change with the Deutsche Gesellschaft für Internationale Zusammenarbeit (GIZ) GmbH in Delhi, India. He advises Indian Ministries and Departments at various levels of government and has implemented projects in the area of waste/resource management, climate change and economic instruments. He previously worked as a Research Fellow at IDS on the circular economy and drivers of green transformations.

Henry Chavez holds a PhD in Social Sciences from the Ecole des Hautes Etudes en Sciences Sociales in Paris, France. He has an interdisciplinary background in social sciences, economics and politics, specialising in social studies of science, technology and innovation; anthropology of global systems; public policy design and evaluation; circular economy; and responsible production and consumption. He has worked as a consultant in the private and public sectors as well as in local social organisations, NGOs and international organisations. He is currently teaching and researching at the University of Ecuador.

An Chen is project manager at TEDA Eco Center, Tianjin, China, where he is responsible for the industrial park's sustainable development and related consulting work. His areas of expertise include waste management, information disclosure, carbon auditing, green-planning and industrial parks assessment. He holds a bachelor's degree in biotechnology from Central South University and a joint master's degree in Environment Technology and Engineering from UNESCO-IHE, University of Chemistry and Technology Prague, and Ghent University.

Victoria Chillcott is a post-graduate from the University of Sussex in Environment, Development and Policy. Her work focuses specifically on the influences on policy regarding plastic pollution in the UK. Victoria's interests also engage with work around sustainability, in particular waste and responsible resource use. In her personal life she is an advocate for using less plastic, buying local produce, ethical consumption and homegrown food.

Peter Desmond is a circular economy strategic advisor supporting firms in the UK and Africa with circular economy planning and implementation. He is the Circular Economy Club Local Organiser for Brighton & Hove. Peter has an MA (Distinction) in Globalisation, Business and Development from the Institute of Development Studies at Sussex University; his dissertation was entitled, 'Towards a circular economy in South Africa – what are the constraints to recycling mobile phones?'. He is co-founder of the African Circular Economy Network, which aims to build a restorative African economy that creates well-being and prosperity for its people whilst regenerating environmental resources.

Timothy J. Foxon is Professor of Sustainability Transitions at SPRU (Science Policy Research Unit), University of Sussex. His research explores the technological and social factors relating to the innovation of new energy technologies, the co-evolution of technologies and institutions for a transition to a sustainable low carbon economy, and relations and interdependencies between energy use and economic growth. He is a member of the UK Energy Research Centre, the ESRC Centre for Climate Change Economics and Policy, the Centre on Innovation and Energy Demand, and the Centre for Research on Energy Demand Solutions. His has published two books, 60 academic articles, 15 book chapters and over 20 research reports. His book on the role of energy in past surges of economic development and the implications for a low carbon transition was published in October 2017.

Claudia Lorena García is an independent consultant at the Americas Sustainable Development Foundation (ASDF). She is a chemical engineer and holds a master's degree in Innovation and Technology Management from the University of Bath, UK. During her master's studies, Claudia researched the opportunities of and challenges to the transition towards a circular economy in Colombia given the country's specific circumstances, such as development gaps in infrastructure and a large informal sector involved in recycling. Lorena continues researching and supporting the circular economy in the American continent is and currently working with ASDF, leading the Circular Economy Platform of the Americas, an initiative that is enabling the circular economy in the Americas through raising awareness and conducting educational and consultancy activities.

Jai Kumar Gaurav is a climate change and environment sector professional with ten years of diverse experience with NGOs, bi-lateral, multi-lateral organisations and the private sector. He completed his masters in Climate Change and Development from the Institute of Development Studies (IDS), University of Sussex. He currently works at GIZ in India. He has previously worked with adelphi, International Development Enterprises India (IDEI), the United Nations Development Program (UNDP) and Sindicatum Carbon Capital on climate change mitigation and adaptation projects. He has also worked on MSW and e-waste management. His research interests are interdisciplinary and relate to political economy, technology and financing.

Jacqueline Gaybor is PhD researcher at the Erasmus University Rotterdam and a lecturer at Erasmus University College in the Netherlands. She has an interdisciplinary background in Law and Development studies with a focus on human rights, gender and conflict studies. She is a specialist in feminist political ecology, care ethics, social studies of science and technology, zero-waste, and responsible production and consumption. She has experience conducting research for international development organisations and on advising governments, policy makers and private sector actors. She has coordinated development projects in different NGOs and has years of experience working in the public sector.

Pragya Gupta is a finance professional with varied skills and interests in development challenges and experience in working with the public, private and NGO sectors. She is passionate about the role of finance and business in development, strengthening public service delivery and community-led initiatives. She is also interested in organisational management and leadership approaches. She has completed an MA in Development Studies from Institute of Development Studies, University of Sussex, and is a qualified charted accountant from the Institute of Chartered Accountants in India.

Jayesh Joshi is Secretary of VAAGDHARA, India. He has more than 20 years of experience in working in the development sector with indigenous communities on a wide range of issues including health, education, livelihoods, agriculture and natural resource management. He has excellent knowledge on issues pertaining to the youth, farmers, women and children of indigenous communities. Mr. Joshi has been providing leadership and technical guidance to VAAGDHARA since 2002. He has also conducted various research studies on the issues related to tribal communities and has represented VAAGDHARA on various national and international platforms, including the United Nations.

Patricia Noble holds a Masters from the Institute of Development Studies at the University of Sussex, with a background in international business, globalisation and development. She is the Director of the Board at the Circular Economy Club, an international non-profit organisation aiming to connect stakeholders and solve challenges collaboratively. Over the past years she has been acutely involved in circular economy research and on the ground initiatives, which has enabled her to present her work at various conferences and speak at the United Nations.

Patrick Schröder is a Research Fellow at the Institute of Development Studies, University of Sussex, UK. His transdisciplinary research concerns the global transition to a circular economy within the context of sustainable consumption and production (SCP) systems and the Sustainable Development Goals (SDGs). Prior to joining IDS, he was based in Beijing from 2008–2015, where he worked extensively in development cooperation programmes of the European Union and Deutsche Gesellschaft für Internationale Zusammenarbeit (GIZ). He holds a PhD in Environmental Studies from Victoria University of Wellington, New Zealand.

Deepak Sharma is a Consultant at VAAGDHARA, India. He has more than 35 years of experience in working with different agencies in the development sector with indigenous communities on a wide range of issues including natural resource management, climate change, livelihoods, sustainable farming and permaculture. He has excellent knowledge on agricultural issues including soil health, nutrition-focused farming and food diversity in indigenous rural communities. He has been providing guidance and leadership on food diversity to various communities since 2002. He has also conducted various research studies on issues related to traditional food and has participated in various national and international platforms.

Yuyang Song is Director of the TEDA Eco Center in Tianjin, China. She founded the centre in 2010 after working as an officer in TEDA Environmental Protection Bureau since 2001. She has extensive experience in promoting international cooperation in the field of energy saving and environmental protection in industrial parks. She is current responsible for establishing a big data environment system in TEDA and promoting sustainable development for industrial parks nationwide. She holds a bachelor's degree in Environmental Management from

Nankai University, Tianjin, and a master's degree in Environmental Planning and Management from Oxford Brookes University.

Atsushi Watabe is a senior researcher at the Institute for Global Environmental Strategies (IGES), Japan. He has worked on issues including rural development, recovery from disaster and transitions to sustainable lifestyles in East Asian and Southeast Asian countries, with a particular interest in the ways people interpret changes in their natural, economic and social environment, and how they organise their day-to-day living. He also coordinates the Sustainable Lifestyles and Education Programme of the United Nations' 10-Year Framework of Programmes on Sustainable Consumption and Production.

ACKNOWLEDGEMENTS

We would like to thank all the participants of the conference 'Sustainable Lifestyles, Livelihoods and the Circular Economy' that was held from 27–29 June 2017 at the Institute of Development Studies (IDS) and the Science Policy Research Unit (SPRU) of the University of Sussex.

Our thanks go to all the members of the Global Research Forum on Sustainable Production and Consumption and the other co-organisers of the event: Tearfund, the SWITCH-Asia Network Facility, SCORAI Europe, the Collaborating Centre on Sustainable Consumption and Production (CSCP), Aalborg University and GRF's MORE Sustainable Lifestyles initiative, in conjunction with the Sustainable Lifestyles and Education Programme of the 10YFP.

Special thanks for their support to the event and this publication go to Richard Gower and Joanne Green (Tearfund), Uwe Weber and Silvia Sartori (SWITCH-Asia Network Facility), Jeffrey Barber and Karen Onthank (Integrative Strategies Forum), Vanessa Timmer (One Earth), Chiara Fratini and Michael Sørensen (Aalborg University) and many more.

Our thanks to Ian Scoones and Nathan Oxley as the editors of this series, and to Hubert Schmitz and Stephen Spratt for their ongoing support of the circular economy topic at IDS.

We would also like to thank the reviewers of the various chapters and the many other researchers and practitioners working in this area who have provided valuable feedback.

Special thanks to Nalini Shekar and her organisation Hasiru Dala Innovations for participating in the conference and for providing the photo for the book cover. Hasiru Dala Innovations (www.hasirudalainnovations.com) is a for-benefit, not-for-loss social enterprise that is focused on creating better livelihoods through inclusive businesses that have an environmental impact. They innovate on circular economy business models that leverage the innate entrepreneurship of waste pickers and in-depth knowledge of waste handling. In collaboration with its sister nonfor-profit, Hasiru Dala (www.hasirudala.in), focusing on securing social justice for waste pickers, it aims to improve the overall quality of life for waste pickers and works with different stakeholders including consumers, producers, government and other waste supply-chain actors to enable a sustainable circular economy.

Finally, we would like to acknowledge the hard work of the millions of waste pickers who make the circular economy happen every day. We hope that through this book we can make a small contribution to support the efforts of improving quality of work and livelihoods of waste pickers.

ABBREVIATIONS

ACEA	African Circular Economy Alliance
ACEN	African Circular Economy Network
CAK	Communications Authority of Kenya
CE	Circular economy
CEAMSE	Ecological Coordination, Metropolitan Area and State Society (Argentina)
CO_2	Carbon dioxide
CSCP	Collaborating Centre on Sustainable Consumption and Production (Germany)
CSE	Centre for Science and Environment
CSO	Civil Society Organisation
CSS	Solidarity Selective Collection (Coleta Seletiva Solidariedade in Portuguese)
DMMT	Disposable menstrual management technologies
DPCC	Delhi Pollution Control Committee
EIP	Eco-industrial park
EMF	Ellen MacArthur Foundation
EPA	Environmental Protection Agency
EPR	Extended producer responsibility
EU	European Union
GAIA	Global Alliance for Incinerator Alternatives
GDKP	Gerakan Diet Kantong Plastik Plastic Bag Diet Movement
GDP	Gross domestic product
GE	Green economy
GESIP	Green Economy Strategy and Implementation Plan (Kenya)
GIZ	Deutsche Gesellschaft für Internationale Zusammenarbeit
GVC	Global value chains

GW	Gigawatt
ICT	Information communication technology
IDS	Institute of Development Studies
IP	Industrial park
IRENA	International Renewable Energy Agency
IWP	Informal waste pickers
MBT	Mechanical biological treatment plant
MSW	Municipal solid waste
MW	Megawatt
NAMA	Nationally Appropriate Mitigation Action
NDMC	New Delhi Municipal Council
NDZ	National Development Zones (China)
NGO	Non-governmental organisation
NISP	National Industrial Symbiosis Programme
OHS	Occupational health and safety
PLA	Participatory learning and action
PPP	Public-private partnership
PSS	Product service systems
PV	Solar photovoltaic
RMMT	Reusable menstrual management technologies
SA	South Africa
SME	Small and medium-sized enterprise
SCP	Sustainable consumption and production
SDGs	Sustainable Development Goals
SHS	Solar home system
SPMA	São Paulo Metropolitan Area
TBNA	Tianjin Binhai New Area
TEC	TEDA Eco Centre
TEDA	Tianjin Economic-Technological Development Area
UN	United Nations
UNICEF	United Nations Children's Fund
UNIDO	United Nations Industrial Development Organisation
WBCSD	World Business Council on Sustainable Development
WEEE	Waste electrical and electronic equipment
WIEGO	Women in Informal Employment: Globalizing and Organizing
WMC	Waste management cycle
WPC	Waste picker cooperative
ZWI	Zero Waste Institute
ZWIA	Zero Waste International Alliance

PREFACE

From 27–29 June 2017, the Institute of Development Studies (IDS) and the Science Policy Research Unit (SPRU) of the University of Sussex cooperated with the Global Research Forum on Sustainable Production and Consumption (GRF) to organise a conference on 'Sustainable lifestyles, livelihoods and the circular economy' in Brighton, UK. This conference paid special attention to the social dimensions of the circular economy (CE), especially examining how the CE concept and approach fares in developing countries and with the poor globally.

The GRF community is composed of both individual researchers and practitioners as well as research and policy organisations and institutions across a wide range of countries and regions, which since the launch in Rio 2012 includes Brazil, China, South Africa, Switzerland, Canada, Peru, Austria, Georgia, Singapore, India, the US and now the UK. The Brighton conference on Sustainable Lifestyles, Livelihoods and the Circular Economy was the product of collaboration amongst a number of organisations. The event was co-organised by Tearfund, the SWITCH-Asia Network Facility, SCORAI Europe, the Collaborating Centre on Sustainable Consumption and Production (CSCP), Aalborg University and GRF's MORE Sustainable Lifestyles initiative, in conjunction with the Sustainable Lifestyles and Education Program of the 10YFP.

The conference built on the ongoing work of the organisers and partners of this conference. In 2016, Tearfund and IDS published the Virtuous Circle report (Gower and Schröder, 2016), which outlines the opportunities the circular economy could offer for low and middle-income countries to improve people's health, generate employment, and protect resources and the environment. The themes of the conference also emphasised the idea that we cannot simply assume that 'models developed for one setting . . . will work in others: whether exported from the developed to the developing world, or from the laboratory or research station to the field' (Leach, Scoones and Stirling, 2007). Therefore, a key goal of the conference was to examine ways in which CE might offer a useful path or tool in achieving sustainability, well-being and social equity in diverse contexts. In particular, the conference explored the CE approach with regard to two other important conceptual frameworks: sustainable lifestyles and sustainable livelihoods – all three viewed in the context of the UN 2030 Agenda – the Sustainable Development Goals (SDGs).

During the three days of presentations and conversations, we explored each of these themes in relation to one another, examining the potential for synergies to address the goals of eradicating poverty and inequality, protecting our environment, and ensuring sustainable production and consumption throughout the world. Our initial call for proposals identified five topic areas:

1. How do we understand the concept and practice of the circular economy in developing countries vs. developed countries, industrialising vs. post-industrialised countries?

2. What are the relationships between sustainable lifestyles and livelihoods and the circular economy?

3. How can transitions towards a more circular economy contribute to the implementation of the Sustainable Development Goals on reducing poverty and promoting sustainable livelihoods?

4. What types of socio-technical innovation, social marketing strategies, public policies and political reforms are required to change lifestyles and the current linear 'throw-away' system to a circular system that operates within planetary boundaries?

5. What types of financial business strategies and supply chains support the transition to a circular economy?

This book is a selection of some of the conference presentations and papers: a representative summary of the breadth of contributions presented during the event. We would like to thank all participants and contributors of the conference. The contributions and discussions of the event go beyond the scope of this book, in which we try to capture the main points of discussions and academic debates that emerged.

References

Gower, R. and Schröder, P. (2016) *Virtuous Circle: how the circular economy can create jobs and save lives in low and middle-income countries*. London and Brighton: Tearfund and Institute of Development Studies.

Leach, M., Scoones, I. and Stirling, A. (2007) *Pathways to Sustainability: an overview of the STEPS Centre approach*. STEPS Approach Paper. Brighton: Institute for Development Studies.

PART I
Introduction

1
INTRODUCTION

Sustainable lifestyles, livelihoods and the circular economy

Patrick Schröder, Manisha Anantharaman, Kartika Anggraeni, Timothy J. Foxon and Jeffrey Barber

Circular economy: a new approach for sustainable development?

Overlapping concepts and domains

In this book we address three separate yet overlapping thematic concepts and domains of knowledge, practice and discourse: sustainable lifestyles, livelihoods and the circular economy. Each of these concepts offers contributions to the overall transition to sustainable production and consumption systems and better lives for all. Common to these different concepts is the desire to provide for human needs and improve the quality of life while reducing social and environmental harm and creating pathways to sustainability (Leach, Scoones and Stirling, 2007).

The concept of *circular economy* (CE) focuses on a set of principles that offer an operational vision of concrete paths to sustainable production and consumption systems and thus to a sustainable economy. The CE approach highlights the importance of changing the current linear model into a system that is regenerative and restorative by design (Ellen MacArthur Foundation, 2015). This can be achieved by redirecting energy and material flows from a linear to a circular direction, transforming waste into productive inputs, and reducing pollution, greenhouse gases and their impacts on health and environment. This involves systems thinking approaches that include changes in value systems, ambitious policies to internalise externalised costs, and new approaches to production, distribution, consumption and investment within each sector of the economy (Stahel, 2016).

Lifestyles is a term used to describe the behavioural codes and cognitive frames enabling decision-making for actions and choices consistent with one's social identity and role within a particular community. This includes roles as consumers of products and services, producers (i.e. workers, managers, shareholders, service providers) and investors. Lifestyles are increasingly complex given that people tend to belong and identify with not one but a cluster of communities. Lifestyles are considered more or less sustainable to the degree that the actions and choices associated with different roles and identities are guided by sustainability values (Leiserowitz, Kates and Parris, 2006).

Livelihoods, in turn, shape lifestyles in relation to the roles people play in acquiring the means of living, whether as construction workers, farmers, professors, managers or artists. The concept of sustainable livelihoods relates to a wide set of issues that encompass the relationships between poverty and environment (Chambers and Conway, 1992). This includes concerns with work and employment, poverty reduction, broader issues of adequacy, security, well-being and capability, and the resilience of livelihoods and the natural resource base on which they depend (Scoones, 1998; Scoones 2015).

The following sections discuss each of these three concepts in more detail, and explore how they relate to each other.

Circular economy definitions – unity in diversity

The circular economy is today a term that means different things to different people. There is a wide range of circular economy thought-schools, including those who associate the term with cradle-to-cradle design, industrial ecology, performance economy, regenerative design and even biomimicry. The roots of the concept of 'circular economy' go back to classical political economists (e.g. Ricardo, Smith, Quesnay) who saw the system of production and consumption as a circular process which 'stands in striking contrast to the view presented by modern theory, of a one-way avenue that leads from "factors of production" to "consumption goods"' (Sraffa, 1960:93). Others cite Kenneth Boulding's 1966 paper 'The economics of the coming spaceship earth' or more recently the work of David Pearce and Kerry Turner (1990) as antecedents of the term. Moving beyond strict adherence to neo-classical economic precepts, CE has been described as a framework for re-designing the economy by the Ellen MacArthur Foundation, which has been championing the concept globally since 2010 (Ellen MacArthur Foundation, 2015).

According to this contemporary school of thought, the CE concept is grounded in the study of non-linear, particularly living systems (Webster, 2017) and refers to an industrial economy that is restorative by design and relies on renewable energy, minimises, tracks, and hopefully eliminates the use of toxic chemicals, and eradicates waste through careful design. Imitating living systems, the CE approach works to optimise systems rather than components (i.e. 'design-to-fit'). This is done through attention to material and energy flows, which according to McDonough and Braungart (2002) can be classified into two kinds: biological nutrients, useful to the biosphere, and technical nutrients, useful to the technosphere, i.e. the systems of industrial production. These definitions of a CE are based on a synthesis of ideas and concepts such as 'cradle to cradle' (McDonough and Braungart, 2002), biomimicry (Benyus, 1997) and the performance/sharing economy (Stahel, 2016), and include insights from industrial ecology. A recent definition by Geissdoerfer et al. (2017), who view the CE as a potential new sustainability paradigm, summarise the main elements of the CE as 'a regenerative system in which resource input and waste, emission, and energy leakage are minimised by slowing, closing, and narrowing material and energy loops. This can be achieved through long-lasting design, maintenance, repair, reuse, remanufacturing, refurbishing, and recycling.'

Overall, there is little consensus and convergence on the definition of the circular economy, and current definitions have a number of limitations (Homrich et al., 2018). Despite the breadth of the concept and related practices of the circular economy, it is becoming increasingly popular among policymakers. Circularity has been adopted as national policy by China in 2009 (see Chapter 8 of this book) and Finland in 2016. The Netherlands adopted a government-wide programme aimed at developing a circular economy in the Netherlands by 2050. According to the Netherlands Environmental Assessment Agency, the approach of making optimal use of raw materials and resources contrasts with the currently dominant linear economy that operates on a 'take-make-dispose' logic, assuming access to unlimited resources and thereby producing products to be discarded after use. A circular economy, on the other hand, 'centres around the reuse of products and raw materials, and the prevention of waste and harmful emissions to soils, water and air, wherever possible', thus closing the loop (PBL, 2017).

In 2015 the European Commission adopted the circular economy concept as part of the EU's 2020 strategy initiative 'to modernise and transform the European economy, shifting it towards a more sustainable direction' (EC, 2015). According to the EU,

> The transition to a more circular economy, where the value of products, materials and resources is maintained in the economy for as long as possible, and the generation of waste minimised, is an essential contribution to the EU's efforts to develop a sustainable, low carbon, resource efficient and competitive economy. Such transition is the opportunity to transform our economy and generate new and sustainable competitive advantages for Europe.

In support of the transition to a circular economy, the EU package includes legislative proposals on waste, with long-term targets to reduce landfilling and increase recycling and reuse. In closing the loop of product lifecycles, the package includes 'an Action Plan to support the circular economy in each step of the value chain – from production to consumption, repair and manufacturing, waste management and secondary raw materials that are fed back into the economy' (EC, 2015).

While there is much excitement about the promise of CE, in assessing the potential for and transition pathways from a linear to a CE, it is important to acknowledge the different capacities, opportunities and pressures at different levels and stages of the process. As Potting et al. (2017) point out,

> the actual circular economy transition should lead to closing cycles at the level of individual products, i.e. in the related product chains. The transition process

may differ across products and between circularity strategies, where lower circularity strategies are still closer to a linear economy and higher circularity strategies are closer to the circular economy.

Technological innovation, they note, is mainly relevant for lower circularity strategies, whereas

> socio-institutional changes become more important for higher circularity strategies increasingly involve transforming the whole product chain (i.e. systemic changes). Socio-institutional changes refer to differences in how consumers relate to products, how all actors in a product chain cooperate to achieve circularity, and all institutional arrangements needed to facilitate this.

Further, researchers have argued that proponents of the circular economy could learn from the social and solidarity economy, as well as from institutional economics, by embedding the CE in relations of power, more explicit value systems and solidarity principles (Moreau et al., 2017). Towards this end, more consideration could be given to the institutional conditions necessary for setting rules that differentiate profitable from non-profitable activities in a circular economy, and guaranteeing high labour standards. Additionally, questions remain as to whether the circular economic model should contribute to 'alternatives to growth', i.e. a sharing rather than for-profit model involving changes to both production and consumption patterns, or simply to an alternative model of growth which would only involve changing forms of production and business models.

Circular economy in the context of green transformations

Despite the limitations, different goals and ongoing conceptual development within the circular economy, there seems to be consensus that the circular economy has, potentially, much to offer in augmenting existing efforts at environmental sustainability and solving global environmental challenges.

Conceptually, in this book, we approach the circular economy through the lens of green transformations, a particular type of thinking about and conceptualising transformations which is primarily concerned with environmental sustainability, but also highlights issues of contested politics and the social dimensions of negotiating pathways. The perspective of green transformations is about 'involving more diverse, emergent and unruly political alignments, more about social innovations, challenging incumbent structures, subject to incommensurable knowledges and pursuing contending (even unknown) end' (Scoones et al, 2015:54). A green transformations perspective emphasises that there is no consensus on the 'drivers' of environmental stress, whether that be overconsumption or rapid urbanisation processes, and even less consensus on the required solutions and processes of change. Simply put, 'a clear vision of what green transformations are required, for what and for whom remains elusive' (Scoones et al., 2015:5).

Furthermore, the green transformations lens highlights the existence of a number of narratives, each reflecting different and sometimes competing framings of environmental problem and solution, and therefore different versions of sustainability. These different narratives also appear in the transformations from a linear system of production and consumption to a circular economy. Research about the circular economy shows that there exist different pathways towards more circularity (Homrich et al., 2017). As we discussed earlier, the concept itself is contested, with several definitions bandied about. Some scholarship on the circular economy aims to achieve consensus about the concept, main principles and approaches (Prieto-Sandoval, Jaca and Ormazabal, 2018). However, we argue that it is not clear if these existing pathways will converge into one main circular economy approach or if they might instead further diverge into multiple pathways characterised by different practices and processes. As Scoones et al. (2015) have pointed out, green transformations will be achieved through a combination of pathways and there is no one-size-fits-all approach, so a diversity of political strategies will be required. In the spirit of the green transformations literature, this book seeks to document the diversity of meanings and practices associated with the circular economy, focusing specifically on the developing world.

Having presented an overview of the history and contemporary understanding of the circular economy and highlighted some of the contestation around what the term means and how it is enacted in practice, the next section discusses the circular economy in relation to narratives around economic development and economic growth.

Economic development narratives and the circular economy

The circular economy focuses on the creation and retention of value associated with natural resources and manufactured products by enhancing circular flows aimed at regenerating and restoring this value. This is usually interpreted as economic value, so that increasing the circularity of flows would enable greater economic value to be created whilst minimising environmental impacts associated with the extraction of natural resources and the emission of wastes. As argued by the Ellen MacArthur Foundation (2015), 'In a circular economy, improving the value captured from existing products and materials, not just increasing their flow, would increasingly drive economic growth'. This is to be achieved by:

- Preserving and enhancing natural capital by controlling finite stocks and balancing renewable resource flows;
- Optimising resource yields by circulating products, components, and materials at the highest utility;
- Fostering system effectiveness by revealing and designing out negative externalities.

Similarly, the EU Action Plan for the Circular Economy emphasises that the maintenance of value of products, materials and resources in the economy for as long as possible is 'an essential contribution to the EU's efforts to develop a sustainable, low carbon, resource efficient and competitive economy' (EC, 2015). Thus, the CE is usually understood in terms of enhancing resource productivity, i.e. the economic value created per unit of resource use, and decoupling economic growth from resource use and environmental impacts.

This means that the CE is usually embedded in a green growth narrative, emphasising new business development and market opportunities, more efficient ways of producing and consuming, and the creation of local jobs. Using a simple economic model, the Ellen MacArthur Foundation (2015) projected that 'GDP could increase as much as 11 percent by 2030 and 27 percent by 2050 in a circular scenario, compared with 4 percent and 15 percent in the current development scenario'. This is argued to be driven by 'increased consumption due to correcting market and regulatory lock-ins that prevent many inherently profitable circular opportunities from materializing' (Ellen MacArthur Foundation, 2015). This is to be achieved, they further argue, whilst reducing CO_2 emissions by 48 percent by 2030 and 83 percent by 2050, compared with 2012 levels.

However, this view creates challenges for mainstream economic theory, in which economic value is interpreted in terms of exchange value, rather than use value. GDP measures value-added in the economy in exchange value terms, i.e. no intrinsic value is assigned to natural resources. Current approaches to measuring natural capital are based on assigning economic (exchange) value, e.g. through contingent valuation, in which people are asked what they would be willing to pay to preserve a particular feature (Farley, 2012). This leads to questions as to whether a green growth perspective, based on mainstream economic theory, will be sufficient to achieve the levels of resource use reduction and carbon emissions reductions needed to mitigate climate change and other environmental impacts.

Firstly, there is a danger that, if value is purely measured in economic exchange terms, then the contribution of resource and energy use to that value creation is neglected, leading to an overestimation of the potential for beneficial resource productivity improvements. If circular economy changes are still embedded in an economy based on ever-increasing levels of consumption, then some or all of the resource productivity improvements associated with CE could be taken back in higher consumption. Secondly, there is a danger that, if the wider value of social benefits associated to circular economy improvements is not adequately measured, the potentially socially beneficial changes will not be implemented. We now briefly explore these dangers.

Research by ecological economists argues that primary energy and resource inputs, together with improvements in the efficiency of their conversion to provide useful work, have been crucial drivers of past surges of economic growth (Ayres and Warr, 2005; 2009; Krausmann et al., 2009; Foxon, 2017). This suggests that at least part of the value embodied in useful goods and services should be attributed to these energy and resource inputs. However, in mainstream economics, no intrinsic value is assigned to these inputs. This means that any energy source or natural resource that can be profitably extracted and sold, will be. Hence, there is no notion of natural limits to these sources. This contrasts with work on planetary boundaries that argues that the overall scale of impact of human economic activities on the planet already exceeds key natural limits, including biodiversity loss, imbalanced nutrient cycles and climate change (Steffen et al., 2015; Raworth, 2017; O'Neill et al., 2018).

As resources are not currently valued, this leads to highly wasteful use of these resources under the current linear economy model, including, for example, short product lives, low utilisation of products and low capture of value at end-of-life. Hence, if barriers to the adoption of CE models can be overcome, then economic value can be created with much lower levels of resource use, thus reducing environmental impacts. However, if this reduces the cost of the final good or service, then there is a danger that these beneficial outcomes of CE could be reduced by increases in overall consumption caused by adoption of these CE models. This is related to the so-called 'rebound effect' associated with energy efficiency improvements, in which these improvements lead to a reduction in the cost of the final energy service, and so to an increase in consumption of that service, reducing or eliminating the expected energy saving (Sorrell, 2015; Korhonen et al., 2018). This suggests that natural resources need to be given an intrinsic value associated with the scarcity of these sources and the ability of the biosphere to assimilate the wastes produced, for example by assigning a value on carbon emissions through imposition of a carbon tax or trading scheme (Baranzini et al., 2017).

Another concern is related to a neglect of the social and institutional dimensions of the CE (Moreau et al., 2017). The implementation of a circular economy requires institutional changes to overcome barriers to implementation, including internalisation of negative externalities such as carbon emissions, regulations to overcome imperfect information on CE alternatives, and education and incentives to overcome cultural lock-in to linear economy business models and user practices. Attention to social dimensions is also important to avoid potential unintended consequences and to ensure that the social benefits of CE solutions are spread widely. For example, the so-called 'sharing economy' has been identified as a key component of CE. This involves the adoption of practices aimed at maximising the utilisation of existing products, often enabled by the application of information and communication technologies. Well-known examples of this include Airbnb, the service that enablers travellers to rent rooms or houses that would otherwise be unoccupied, and Uber, the ride-hailing service that connects users wanting a ride with car owners willing to drive them. Whilst proving highly popular with users, these services have be criticised for creating new incentives to travel, crowding out existing providers that may provide a higher value service, and monetising previously freely shared services as well as offering low job security and in-work benefits to providers (Frenken and Schor, 2017). This suggests that the wider social benefits and potential unintended consequences of CE solutions need to be considered in designing institutional changes and incentives to promote their adoption.

Whilst still recognising the high potential of CE solutions to deliver economic value to users and businesses through significantly reducing resource use and environmental impacts, the recognition of these dangers leads to a note of caution and recommendation not to see CE as a panacea for solving all environmental problems. CE solutions need to be embedded in an institutional context that recognises planetary boundaries for the overall scale of human consumption and production activities, and seeks to design systems that promote wider social well-being. In other words, in order to achieve a CE pathway, social, political, institutional negotiations within planetary boundaries must take place. This means going beyond a technical-economic framing of CE towards one that is centrally about the politics of sustainability and development (Scoones, 2016). The following section delves into the politics of sustainability and development as relevant to the circular economy, focusing explicitly on North–South power relations.

Circular economy in the context of international development and North–South power relations

Green industrial development and global value chains

The circular economy concept not only offers potentials for environmental sustainability in the Global North, but could also offer inspiration for new development models that support economic and social objectives in the Global South. According to a report by Chatham House, CE could help to resolve some of the dilemmas posed by business-as-usual development models

> by increasing economic productivity, generating employment and reducing exposure to volatility in raw materials prices. At the same time, CE strategies could avert some of the major pressures facing developing countries – including health and environmental effects from unmanaged waste – with clear benefits in terms of lives saved as a result of reduced air, water and soil pollution.
> *(Preston and Lehne, 2017:6)*

Countries in the Global South are beginning to realise the opportunities the circular economy might offer for development, especially for greening industrial development and addressing challenges of increasing waste generation. The CE concept might be best articulated as a green industrialisation strategy that can avoid negative externalities to environment and helps safeguard development gains (Preston and Lehne, 2017). In this context it links to and supports existing green economy approaches and industrial development objectives, but goes beyond limited resource efficiency objectives and opens up new perspectives on the future of green industrial strategies. For example, Colombia is promoting CE as an industrial development strategy and encouraging businesses towards this change (see Chapter 7 of this book), while multilateral development organisations such as UNIDO are promoting CE approaches such as industrial symbiosis in a number of countries including Uruguay and Vietnam. In African countries, industrial repair and remanufacturing industries such as the Suame/Kumasi automotive cluster (Schmitz, 2015) have been operating successfully for decades. More recently, in November

2017 the World Economic Forum and the Global Environment Facility established the Africa Circular Economy Alliance (ACEA) in collaboration with the governments of South Africa, Nigeria and Rwanda (Molewa, 2017) (see Chapter 9).

However, to date, only few studies have explored the potentials and demand for CE approaches among stakeholders in poorer countries (see, for example, Gower and Schröder, 2016), or have sought to understand where the CE concept and narrative might complement or conflict with existing priorities and pathways.

Narratives and power relations

To understand how conceptions and narratives of the CE might differ in the developing vs. developed world, we must first examine these categories themselves. The practice of identifying certain countries as developed and others as developing is both ubiquitous and heavily debated. For example, these broad categories have been critiqued by scholarly interventions that question or reject the very notion of development and the binaries it suggests (Escobar 1995; Sparke 2007). Similarly, development agencies such as the World Bank have moved away from the use of the binary given the heterogeneity between different countries classified in the same group. Nevertheless, these simplified categories (developing vs. developed or North vs. South) help us examine geographic and ideological diversity within the CE. In this volume, we rely on this binary to highlight the ways in which CE principles are being revived or innovated in the developing world, and through this provide insights as to how the concept might be reimagined from the positionality of the South.

Thinking about what the CE might look like from the perspective of the South is important because mainstream notions of development and sustainability have been repeatedly challenged by voices in the margin. Questions that could be raised in this regard include looking at the politics of identity and power involved in deciding what counts as circular, or what is a sustainable lifestyle or livelihood (see Chapter 2). While the CE represents a challenge to the conventional linear economic paradigm, there is the question as to what additional perspectives, particularly from the South, may challenge and hopefully improve the concept and practice of CE as it currently stands. Conceptual expansion driven by ideas from the South has already been witnessed, for instance, in discussions around the concept of sustainable development. As Najam (2005) notes, in the domain of sustainable development, countries of the Global South moved from initial scepticism and reluctant engagement to active agenda-setting, a pattern that is now being seen in CE as well, as discussed in the following.

What is often overlooked are the dynamics and politics of relationships between governments, businesses and people in the North vs. in the South, all of which are implicated in every aspect of CE. Value chains and waste cycles have been global for some time now. Waste in particular has been a transnational issue marked by contestation and negotiation between the North and the South. From the crises around toxic dumping in the Global South that prompted the creation of the Basel Convention to China's new agenda of restricting the import of 'foreign garbage' from the west (see Chapter 3), unequal North–South power relations characterise the functioning of the existing (circular) economy. While historically, countries in the Global North seemed to hold all the power in the domain of transnational waste dumping and trade, the situation is rapidly evolving, with several countries in the Global South instituting domestic policies to protect local environments. In 2018, China banned waste imports as the country is waging a 'war on pollution' (Reuters, 2018). Before that, in July 2016, east African countries like Rwanda, Kenya, Tanzania and Uganda decided to increase tariffs on imported second-hand textiles as cheap clothes from abroad were threatening local industries (CNBC, 2018).

Responding to pressure from countries in the South, economies in the North are having to rethink domestic policies and create novel collaborations. For example, the EU was working together with four African countries to develop e-waste recycling facilities (EC, 2015), since most of e-waste from EU countries ends up in Africa and Asia (see, for example, Imran et al., 2017). As the situation of transnational waste dumping and trade demonstrates, CE cannot be achieved in a vacuum, but only through collaboration between countries. Transferring waste (or responsibility) to other countries is not possible in the long run as this would simply mean moving pollutants to 'other corners' of our world, an unjust solution in the short term, and an unsustainable one in the longer term.

Beyond the issue of waste, we also see unequal power relations playing out in global value chains that are largely controlled by powerful business actors located in the Global North. Higher-value employment-generating opportunities like product design, marketing and retail are sited in the North, while activity with lower economic value revolving around sorting, reusing and recycling waste is relegated to lower-income countries. For an inclusive transformation to a CE on the planetary scale, we cannot overlook these systemic issues of unequal power relations entrenched in global value chains (Schröder et al., 2018). All these issues call for more critical and systemic approaches to theorising and imagining a circular economy.

Sustainable livelihoods and lifestyles in the circular economy

While the concept of CE provides a clear picture of how waste and emissions can be reduced through a redirection of energy and material flows, it may not be as clear on how this approach generates sustainable livelihoods for whom in the context of the unequal patterns of use and ownership located in a contested political economy (Scoones, 2015). Thus, one of the key questions raised in this edited volume is to whether and how CE practices generate sustainable livelihoods. We provide brief explanations on each of these terms before discussing them in relation to the CE.

The idea of 'sustainable livelihoods', an attempt to go beyond conventional definitions and approaches to eradicating poverty, was introduced in 1987 by the Brundtland Commission on Environment and Development, followed by the 1992 UN Conference on Environment and Development (Krantz, 2001). One of the more frequently cited definitions of sustainable livelihoods comes from Chambers and Conway (1992), who explain

> a livelihood comprises the capabilities, assets (stores, resources, claims and access) and activities required for a means of living: a livelihood is sustainable which can cope with and recover from stress and shocks, maintain or enhance its capabilities and assets, and provide sustainable livelihood opportunities for the next generation; and which contributes net benefits to their livelihoods at the local and global levels and in the short and long term.

While CE efforts highlight the possibility of new industrial service jobs in recovery, reuse, repair, remanufacturing and recycling, there remains the question of the reduction of jobs following declines in new product sales and loss of manufacturing jobs resulting from increased sharing, self-service technologies and extended product life and use. While WRAP offers an optimistic view with new CE job opportunities across all skill levels reducing the 'occupational mismatch' in high unemployment regions, such as the north of England (Morgan and Mitchell, 2015), to what degree circularity can ensure not just more jobs but more sustainable livelihoods is still an open question. Further, the question remains as to who will benefit from these job opportunities.

In the specific context of sustainable rural livelihoods, the CE concept appears to support and overlap with many traditional forms of farming, organic agriculture and restorative practices which benefit smallholder livelihoods. Examples which highlight this link and the synergies between circular agricultural practices and sustainable livelihoods are traditional rural communities in India (Chapter 10) and organic farming practices in Thailand (Chapter 11).

In addition to the concept of sustainable livelihoods, we must also consider the issue of what types of sustainable lifestyle the CE can provision. The origin of the term 'lifestyles' is often attributed to the sociologists Thorstein Veblen and Max Weber and the psychologist Alfred Adler. Both Veblen and Weber noted the role of the lifestyle as a public expression of status, while Veblen (1899) especially associated with the concept of 'conspicuous consumption'. On the other hand, Giddens (1991) views lifestyle as a complex of social practices but also as a strategic vehicle, which individuals use to negotiate a complex diversity of options in navigating modern social life, in contrast to the rigid behavioural rules and markers assigned by traditional society to designate and enforce class identity, role and expectations.

Sustainable lifestyles are therefore not simply about affluent consumers consuming less. Overall, they are life strategies that strive to cause the least harm to people and environment. In this sense, one's livelihood is an essential component of a lifestyle, a critical determinant as to the opportunities and capacities enabling a lifestyle to take shape and function. In turn, a lifestyle can become the means to cultivate, maintain and expand livelihood opportunities and capacities, such as becoming a union member or collective. The same applies to participation in circular vs. linear economy practices and processes, whether as employee, consumer, manager or investor. The idea implied here is that participation in a CE would require shifts in lifestyle patterns that go beyond a single domain to encompass larger life choices.

Constricted by poverty and limited access to resources and opportunity, lifestyles of the poor have also shown resilience, resourcefulness and innovativeness in ways the affluent could do well to learn. However, the overall trend globally has been towards less and less sustainable lifestyles. As increasing numbers in developing countries move into the new consumer classes, so are the consumption habits of the affluent increasingly adopted – larger houses, meat and dairy consumption, car ownership and air travel. While rising standards of living is a goal, the negative impact on health and environment is not. Thus considering the question of lifestyles explicitly within a CE framework is critical at this juncture of intensifying consumption.

Inside and outside the circle: is sustainability a luxury?

Some will point out that these circular/sustainability practices imply more affluent consumers who have access to sufficient resources and assets, allowing them the freedom to engage in 'alternative means of satisfying needs' in contrast to the poor, who are engaged in the daily struggle to simply survive. As Auma Obama declared, 'sustainable living is a luxury . . . the question is: Who can afford that?' (Deutsche Welle, 2012), noting that 'green economy is a western term meaning that the economy has to become more environmentally friendly. However, often that doesn't come up in African countries'.

While lifestyle strategies are about individual choice, we need to take into account how the range of options, opportunities and abilities constrict or enable the choice of sustainable practices and lifestyles. Campaigns to 'change lifestyles' might be more effectively framed in terms of expanding the freedom to choose a sustainable pathway, by paying attention to enhancing the abilities and opportunities enabling sustainable practices and lifestyles. While this includes 'having the relevant information and awareness', it also means establishing the policies and infrastructure changes needed to overcome the various barriers that impede motivation as well as effective behaviour.

Who should these strategies and programmes promoting circular/sustainable practices and lifestyles aim to reach? Are they only appropriate for the affluent? How should the question be reframed in a way that will make it more relevant and appropriate for the poor? If the focus is primarily on reducing carbon emissions, then the target audience would more likely be the more affluent and over-consuming populations. However, if the focus is on meeting needs and improving well-being while minimising harm to others and the environment, then the sustainable livelihood approach is especially relevant, highlighting those capabilities and opportunities needed to enable a truly sustainable lifestyle. A CE that leaves out the poor is not completing the circle. The world cannot afford making sustainability a luxury.

Circular economy: marginalisation or inclusive development?

If the CE holds the promise of creating new jobs and livelihoods, while also reducing environmental degradation, then the question remains as to who exactly will benefit from these opportunities. Historically, in the context of sustainable development, failure to explicitly consider the needs and lived experiences of marginalised populations (women, indigenous people, etc.) has resulted in poor development outcomes. Thus, in the CE, policymakers, practitioners and thought-leaders might be well-advised to think about existing patterns of inequality and how these might be ameliorated or exacerbated in the CE.

One specific population that deserves mention in relation to CE are waste pickers and other members of the informal waste sector. In many cities across the Global South, thousands of individual waste pickers, sometimes organised as cooperatives, sustain themselves by extracting recyclable waste from street dumps and landfills, diverting these to recycling value chains (Wilson, Velis and Cheeseman, 2006). The work of waste pickers, scrap dealers and informal sector recyclers represents an existing form of the circularity, which has increasingly come under threat. In several cities in Asia and Latin America, attempts at modernising waste collection and disposal systems has resulted in the displacement of waste pickers, many of whom are women, indigenous or otherwise marginalised peoples. Organisations like WIEGO are now advocating for the formal inclusion of waste pickers and informal sector recyclers in the newly emerging CE, arguing that work opportunities in this sector should be made available to existing waste picking populations.

However, as scholars have argued, the politics of inclusion is complicated by itself (Anantharaman, 2014). For example, as Reddy (2015) shows, improvement schemes designed by development agencies effectively marginalised informal sector e-waste recyclers in Bangalore as abject residents who were restricted to collecting and manually processing waste from the most marginal frontiers of the city's e-waste circuits. What this suggests is the following: even when waste pickers and other informal sector recyclers are included in circular economy schemes, they are included solely in execution roles or framed as beneficiaries of schemes. The experiential knowledge and innovative entrepreneurialism of these groups are frequently ignored. This raises the question of who is seen as a legitimate knowledge creator and expert in the CE, and one of the contributions of this edited volume lies in identifying case studies of existing circular practices in the Global South that could serve as models elsewhere.

Issues of gender equity also deserve special attention in the context of the CE. As several scholars have argued, development projects that have failed to problematise the issue of gender exclude and marginalise women, depriving them of livelihoods and status (Kabeer, 1994; Marchand and Parpart, 1995). Additionally, there has been a tendency by scholars in the west to homogenise the experiences of women in developing nations, and this failure to understand the diversity of experiences and situations that women find themselves in results in prescriptions that are not suitable for the context. Paying specific attention to the ways in which CE initiatives can contribute to improving the lives of women living in poverty, or at the very least not exacerbate gender inequity, should be a necessary part of the CE agenda. The SDGs and the CE could potentially open new ways for women to achieve economic sovereignty and livelihood security via new home-based production opportunities or work in waste collection and upcycling. In the chapters on women waste pickers in Brazil (Chapter 4), women's empowerment through recent CE initiatives in Indonesia (Chapter 5) and women farmers in central India (Chapter 10), we see some examples of how this could be realised.

Introduction to the book chapters

The contributions of this book aim to broaden and advance the conversations on the circular economy by explicitly examining the diversity of practices that occur under this umbrella concept, and through this highlight some of the conceptual and practical tensions that exist in the CE.

In Part II of the book, the authors explore the different narratives and politics around municipal solid waste, especially plastics, lifestyles and livelihoods, highlighting issues of inequality and tensions between formal and informal systems.

In Chapter 2, Ashish Chaturvedi, Jai Kumar Gaurav and Pragya Gupta explore the various narratives of the CE in the context of Indian cities. The focus of the existing literature on CE has largely been techno-managerial with a large emphasis on the role of big business in solving the problem. However, as the chapter shows, a large part of resource and waste flows in India happen outside the formal economy. The question is how the techno-managerial solutions will impact on the vulnerable and marginal communities in urban areas and how the multiple contestations in the material and discursive arenas will play out as the circular economy advances. The authors highlight the limited focus on the politics of these flows in the existing CE literature by applying an actor-based conceptual framework that combines circular economy models and situates them in the variable power geometries of actors involved in these material flows in urban areas in India.

The politics of marine plastics pollution are explored in Chapter 3. Patrick Schröder and Victoria Chillcott highlight the politics behind the global trade in plastics waste, the most prominent example being China's ban on importing 'foreign garbage' from the West. The chapter also tries to unpack issues around consumer politics, recently announced corporate 'zero plastics' initiatives and ongoing lobbying against hard policies and bans of certain plastic products. The political economy of closing land-based leakage points of global and local plastic value chains and questions about sources of plastic waste will be crucial to solve marine plastics pollution.

In Chapter 4, Patricia Noble unpacks the contributions of informal waste pickers to closing the waste management cycle in the megacities of Delhi in India and São Paulo Municipal Area (SPMA) in Brazil. The chapter identifies the conditions in both cases that enable informal waste pickers to contribute to improved recycling rates and how waste picker cooperatives contribute to improving livelihoods of the urban poor. The findings are further analysed through a green transformation lens, to understand the specific conditions that catalyse informal waste pickers to recycle and close waste management cycles, which will be crucial to support the wider transformation to a circular economy.

Chapter 5 on women and CE initiatives in Indonesia by Priliantina Bebasari shows the important role that women have today in the fight against plastic waste and in achieving a circular economy. The chapter identifies circular economy initiatives in Indonesia which have provided poor local women with new knowledge and skills in the upcycling of non-hazardous solid waste to obtain new sources of income to support their families. Not only that, some women have even had the chance to taste a more political life as they were involved in some advocacy work and were later invited by the local governments to contribute in formulating local policy on waste management. More systematic support from the national government is required so that such women empowerment efforts do not happen randomly and in the short term or lack a clear direction and long-lasting impacts. In the chapter it also becomes clear that CE is a new concept, though the upcycling activities were implemented long before the CE concept 'arrived' in Indonesia.

The Argentinean Zero Waste Framework is analysed in Chapter 6 by Jacqueline Gaybor and Henry Chavez. The chapter discusses and analyses the progress and limitations of a particular case in Buenos Aires, Argentina, that is in the process of implementing a zero waste framework. Although waste prevention and re-use are considered guiding principles of the framework, neither have been prioritised in practice. Instead, the local waste management has been characterised by end-of-pipe solutions which include recycling and landfilling. This paper discusses the management of waste from disposable menstrual technologies and seeks to highlight the importance of prevention and re-use.

Part III of the book highlights the emerging relationship of the circular economy to national industrial policy frameworks and business approaches.

Chapter 7 by Claudia Garcia and Steve Cayzer provides an assessment of Colombia's circular economy transition readiness at a national level. The circular economy offers an alternative development narrative for Columbia which promotes prosperity while regenerating natural capital via innovation and new business models as opposed to current natural resource extraction strategies. This narrative contains valuable insights for Colombian development, while the Colombian context provides a useful perspective on mainstream euro-centric circular economy narratives. This chapter explores the enablers that would facilitate the transition towards a CE in Colombia given its specific circumstances. An enabling framework is proposed to assess the current state of CE in Colombia and to identify the main interventions that are required to support a transition towards a more circularity of production and consumption systems. The assessment shows that Colombia does not yet have the right enabling conditions for a circular economy, but several opportunities are identified, including greater political coherence, a suitable fiscal framework for sustainable practices, a robust IT infrastructure, and use of ICT by enterprises to develop CE business models. The findings of this chapter are specific to Colombia but have relevance for CE transitions in other low and middle-income economies.

In Chapter 8, An Chen, Yuyan Song and Kartika Anggraeni take a look at China's CE and industrial park developments, where industrial symbiosis has become an important strategy. The specific case study of the Tianjin TEDA Eco Center highlights the role of capacity building and the contribution that international development cooperation can make to promoting circular economy practices in green industrial development. Furthermore, the case study also shows the importance of stakeholder engagement, trust building and facilitation of partnerships between companies and local governments to make industrial symbiosis work in China's eco-industrial parks. The chapter also highlights remaining challenges that need to be overcome, including appropriate resource pricing and fee systems for waste to fully reflect environmental impacts and to factor in resource scarcity concerns.

In Chapter 9 Peter Desmond and Milcah Asamba explore the current state of the circular economy in Africa, where the circular economy as a concept is still vague with case studies remaining largely hidden. The legal and regulatory frameworks needed to foster circularity are still in their infancy in most African countries and mechanisms to realise the transition towards green economies are often not in place. Looking at circular economy policies and practices in Kenya and South Africa in the sectors of renewable energy equipment and electronic devices, the chapter explores how the transition towards a circular economy in Africa can contribute towards the achievement of the UN Sustainable Development Goals through the creation of national and regional roadmaps. Opportunities are explored to apply sustainable principles and strategies in a variety of contexts to benefit economies, livelihoods and the environment.

Part IV explores the evolving relationship between rural livelihoods, traditional circular economy practices in agriculture and changing lifestyles of farming communities.

In Chapter 10 Deepak Sharma and Jayesh Joshi provide insights into traditional circular economy approaches among endogenous rural communities in South Rajasthan. The chapter shows how circular economy principles are interlinked with nutrition-sensitive agriculture, how it can be utilised to address malnutrition among endogenous communities of India and how to address Sustainable Development Goal 2 'Zero Hunger'. It highlights the work and research of the civil society organisation VAAGDHARA, applying principles of the circular economy for the revival of the traditional nutrition sensitive farming system with endogenous communities in India. Participatory tools were developed by different agencies and participatory learning and action (PLA) tools were customised and applied across 30 villages of the Banswara districts. Based on this research, the authors developed a framework to apply approaches of traditional circular economy principles in agriculture and support sustainable lifestyles.

The concept of the Thai Sufficiency Economy and how it links with lifestyles and the circular economy is explored in Chapter 11 by Atsushi Watabe. In the chapter, organic farming practices are explored as an opportunity to encourage people to adopt a sufficiency economy as a contribution to the shift of the agricultural sector to the circular economy. The case study of a peri-urban village in Northeast Thailand reveals that people have differing attitudes and contrasting narratives to organic farming, derived from varying livelihood conditions and concerns and lifestyle aspirations of farmers to maintain their living going into the future.

Part V is the concluding section of the book and will provide a brief outlook for the circular economy in developing countries. In Chapter 12, the book's editors Patrick Schröder, Manisha Anantharaman, Kartika Anggraeni and Tim Foxon provide an outlook on the circular economy potentials for the SDGs and international development cooperation, while also highlighting potential risks and pitfalls that the circular economy could pose to developing countries.

References

Anantharaman, M. (2014) Networked ecological citizenship, the new middle classes and the provisioning of sustainable waste management in Bangalore, India. *Journal of Cleaner Production, Special Volume: sustainable production, consumption and livelihoods: global and regional research perspectives*, 63, 173–183. [Online] Available at: https://doi.org/10.1016/j.jclepro.2013.08.041.

Ayres, R.U. and B. Warr (2005) Accounting for growth: the role of physical work. *Structural Change and Economic Dynamics*, 16 (2), 181–209.

Ayres, R.U. and B. Warr (2009) *The Economic Growth Engine: how energy and work drive material prosperity*. Cheltenham and Northampton, MA: Edward Elgar.

Baranzini, A., van den Bergh, J., Carattini, S., Howarth, R., Padilla, E. and Roca, J. (2017) Carbon pricing in climate policy: seven reasons, complementary instruments, and political economy considerations. *WIREs Climate Change*, 2017, 8, e462.

Benyus, J. (1997) *Biomimicry: innovation inspired by nature*. NY: William Morrow.

Chambers, R. and Conway, G. (1992) *Sustainable Rural Livelihoods: practical concepts for the 21st century*. IDS Discussion Paper No. 296. Brighton, UK: Institute of Development Studies.

CNBC (2018) The US has another trade spat going on – and it's over old clothes. [Online] Available at: www.cnbc.com/2018/05/30/us-and-rwanda-trade-dispute-over-second-hand-clothes.html [Accessed 31 August 2018].

Deutsche Welle (2012) Africa Expert: 'sustainable living is a luxury'. *Deutsche Welle*, 30 November 2012. [Online] Available at: www.dw.com/en/africa-expert-sustainable-living-is-a-luxury/a-16418370.

EC (2015a) *Closing the Loop: an EU action plan for the circular economy*. Brussels: European Commission.

EC (2015b) EU-Africa Synergies for e-waste recycling. [Online] Available at: https://ec.europa.eu/programmes/horizon2020/en/news/eu-africa-synergies-e-waste-recycling [Accessed 31 August 2018].

Ellen MacArthur Foundation (2015) *Growth Within: a circular economy vision for a competitive Europe*. Ellen MacArthur Foundation, SUN and McKinsey Centre for Business and Environment.

Escobar, A. (1995) *Encountering Development: the making and unmaking of the Third World*. Princeton, NJ: Princeton University Press.

Farley, J. (2012) Ecosystem services: the economics debate. *Ecosystem Services*, 1 (1), 40–49.

Foxon, T.J. (2017) *Energy and Economic Growth: why we need a new pathway to prosperity*. Abingdon and New York:Routledge.

Frenken, K. and Schor, J. (2017) Putting the sharing economy into perspective. *Environmental Innovation and Societal Transitions*, 23, 3–10.

Geissdoerfer, M., Savaget, P., Bocken, N. and Hultink, E. (2017) The circular economy: a new sustainability paradigm? *Journal of Cleaner Production*, 143 (2017), 757–768.

Giddens, A. (1991) *Modernity and Self-Identity: self and society in the late modern age*. Cambridge, UK:Polity Press.

Gower, R. and Schröder, P. (2016) *Virtuous Circle: how the circular economy can create jobs and save lives in low and middle-income countries*. London and Brighton: Tearfund and Institute of Development Studies.

Homrich, A.S., Galvão, G., Abadia, L.G. and Carvalho, M.M. (2018) The circular economy umbrella: trends and gaps on integrating pathways. *Journal of Cleaner Production*, 175, 525–543.

Imran, M., Haydar, S., Kim, J., Awan, M. and Bhatti, A. (2017) E-waste flows, resource recovery and improvement of legal framework in Pakistan. *Resource, Conservation and Recycling*, 125, 131–138.

Kabeer, N. (1994) *Reversed Realities: gender hierarchies in development thought*. London and New York: Verso.

Korhonen, J., Honkasalo, A. and Seppälä, J. (2018) Circular economy: the concept and its limitations. *Ecological Economics*, 143, 37–46.

Krantz, L. (2001) *The Sustainable Livelihood Approach to Poverty Reduction: an introduction*. Stockholm:Swedish International Development Cooperation Agency.

Krausmann, F., Gingrich, S., Eisenmenger, N., Erb, K.-H., Haberl, H. and Fischer-Kowalski, M. (2009) Growth in global materials use, GDP and population during the 20th century. *Ecological Economics*, 68 (10), 2696–2705.

Leach, M., Scoones, I. and Stirling, A. (2007) *Pathways to Sustainability: an overview of the STEPS Centre approach*. STEPS Approach Paper. Brighton: Institute of Development Studies.

Leiserowitz, A., Kates, R. and Parris, T. (2006) Sustainability values, attitudes, and behaviors: a review of multinational and global trends. *Annual Review of Environment and Resources*, 31.

Marchand, M.H. and Parpart, J.L. (eds.) (1995) *Feminism/Postmodernism/Development*. First edition. London and New York: Routledge.

McDonough, W. and Braungart, M. (2002) *Cradle to Cradle: remaking the way we make things*. New York: Northpoint Press.

Molewa, E. (2017) South African Government Environmental Affairs Minister speech in Bonn, Germany, 11 November 2017. [Online] Available at: www.environment.gov.za/sp eech/molewa_cop23africaalliance_circular_economylaunch [Accessed 22 February 2018].

Moreau, V., Sahakian, M., Griethuysen, P.v. and Vuille, F. (2017) Coming full circle: why social and institutional dimensions matter for the circular economy. *Journal of Industrial Ecology*, 21 (3), 497–506.

Morgan, J. and Mitchell, P. (2015) *Opportunities to Tackle Britain's Labour Market Challenges through Growth in the Circular Economy*. London: WRAP/Green Alliance.

Najam, A., (2005) Developing countries and global environmental governance: from contestation to participation to engagement. *International Environmental Agreements*, 5, 303–321. [Online] Available at: https://doi.org/10.1007/s10784-005-3807-6

O'Neill, D., Fanning, A., Lamb, W. and Steinberger, J. (2018) A good life for all within planetary boundaries. *Nature Sustainability*, 1, 88–95.

PBL (2017) *Food for the Circular Economy*. PBL Policy Brief. PBL Netherlands Environmental Assessment Agency.
Potting, J., Nierhoff, N., Montevecchi, F., Antikainen, R., Colgan, S., Hauser, A., Günther, J., Wuttke, J., Jørgensen Kjær, B. and HanemaaijerA. (2017) Input to the European Commission from European EPAs about monitoring progress of the transition towards a circular economy in the European Union. European Network of the Heads of Environment Protection Agencies (EPA Network) – Interest Group on Green and Circular Economy.
Preston, F. and Lehne, J. (2017) *A Wider Circle? The circular economy in developing countries*. Briefing December 2017. London: Chatham House.
Prieto-Sandoval, V., Jaca, C. and Ormazabal, M. (2018) Towards a consensus on the circular economy. *Journal of Cleaner Production*, 179 (2018), 605–615.
Raworth, K. (2017) *Doughnut Economics: seven ways to think like a 21st-century economist*. London: Random House.
Reddy, R.N. (2015) Producing abjection: e-waste improvement schemes and informal recyclers of Bangalore. *Geoforum*, 62, 166–174. [Online] Available at: https://doi.org/10.1016/j.geoforum.2015.04.003
Reuters (2018) China adds 16 new scrap products to banned import list. [Online] Available at: www.reuters.com/article/china-waste-imports/update-1-china-bans-imports-of-16-more-scrap-waste-products-from-end-2018-ministry-idUSL3N1RW1UK [Accessed 31 August 2018].
Shepherd, C. (2015) The role of women in international conflict resolution. *Hamline University's School of Law's Journal of Public Law and Policy* [e-journal], 36 (2), 53–67. [Online] Available at: http://digitalcommons.hamline.edu/jplp/vol36/iss2/1 [Accessed 30 August 2018].
Schmitz, H. (2015) *Africa's Biggest Recycling Hub?* IDS Blog. Brighton: Institute of Development Studies. [Online] Available at: www.ids.ac.uk/opinion/africa-s-biggest-recycling-hub
Schröder, P., Dewick, P., Kusi-Sarpong, S. and Hofstetter, J. (2018) Circular economy and power relations in global value chains: tensions and trade-offs for lower income countries. *Resources, Conservation and Recycling*, 136 (September 2018), 77–78.
Scoones, I. (1998) *Sustainable Rural Livelihoods: a framework for analysis*. IDS Working Paper 72. Brighton, UK: Institute of Development Studies.
Scoones, I. (2015) *Sustainable Rural Livelihoods and Rural Development*. UK: Practical Action Publishing and Winnipeg, CA: Fernwood Publishing.
Sparke, M. (2007) Everywhere but always somewhere: critical geographies of the Global South. *The Global South*, 1 (1), 117–126.
Stahel, W. (2016) The circular economy. *Nature*, 531, 435–438. [Online] Available at: doi:10.1038/531435a
Sorrell, S. (2015) Reducing energy demand: a review of issues, challenges and approaches. *Renewable and Sustainable Energy Reviews*, 47, 74–82.
Sraffa, P. (1960) *Production of Commodities by Means of Commodities*. Cambridge, UK:Cambridge University Press.
Steffen, W., Richardson, K., Rockström, J., Cornell, S., Fetzer, I., Bennett, E., Biggs, R., Carpenter, S., de Vries, W., de Wit, C., Folke, C., Gerten, D., Heinke, J., Mace, G., Persson, L., Ramanathan, V., Reyers, B. and Sörlin, S. (2015) Planetary boundaries: guiding human development on a changing planet. *Science*, 347, 1259855.
UNESCO (2010) *Global Trend Towards Urbanisation*. [Online] Available at: www.unesco.org/education/tlsf/mods/theme_c/popups/mod13t01s009.html [Accessed 30 August 2018].
UN Women (2016) *Women and Sustainable Development Goals*. [Online]. Available at: https://sustainabledevelopment.un.org/content/documents/2322UN%20Women%20Analysis%20on%20Women%20and%20SDGs.pdf [Accessed 30 August 2018].

Veblen, T. (1899) *The Theory of the Leisure Class*. New York: New American Library.

Webster, K. (2017) *The Circular Economy: a wealth of flows*. Isle of Wight: Ellen MacArthur Foundation.

Wilson, D.C., Velis, C. and Cheeseman, C. (2006) Role of informal sector recycling in waste management in developing countries. *Habitat International, Solid Waste Management as if People Matter*, 30, 797–808. [Online] Available at: https://doi.org/10.1016/j.habitaint.2005.09.005

PART II
Narratives and politics of waste and the circular economy in the Global South

2
THE MANY CIRCUITS OF A CIRCULAR ECONOMY[1]

Ashish Chaturvedi, Jai Kumar Gaurav and Pragya Gupta

Introduction

Changing metabolism in Indian cities

Rapid urbanisation in India has led to the development of large urban agglomerations with a dense concentration of population and economic activity, which causes significant environmental stresses. The stresses arise because cities act both as sinks of resources and as generators of environmental degradation (Kumar et al., 2017; Imran et al., 2017). The provisioning of resources needed by urban inhabitants creates environmental degradation at the origin of these flows of resources; furthermore, significant levels of effluents and discharges are caused that lead to environmental degradation within urban boundaries and beyond.

The first characteristic of such flows is that they are spatially distributed within and beyond city boundaries. This inflow and outflow of resources suggest that cities are open systems with resources flowing in and out of the city boundaries. However, while cities draw the benefits from the inflows of resources, significant damage is caused by the outflows.

The second characteristic of such flows is that they are embedded in contestations in both material and discursive arenas. Actors with several often-conflicting interests are affected by and involved in managing these flows of resources. The governance of these flows is critical to ensure that the most vulnerable actors involved in these flows have equitable access to resources and are not disproportionately over-burdened by the degradation caused by such flows. The current trajectory of the urban metabolisms of large cities in India is clearly unsustainable, with its impact felt within cities as well as in peri-urban areas. It is clear that sustainable urbanisation would need to take into account the spatial distribution of these flows of materials as well as the contestations of these flows in the material and discursive arenas.

Closing material cycles through the circular economy

Over the last few years, the idea of a circular economy (CE), driven by narratives of material scarcity and enhanced resource efficiency, has been gaining traction around the globe.

A large part of the existing literature on the CE has largely been techno-managerial with an emphasis on the role of big business in solving the problem. Further, the existing literature largely focuses on the technical definitions of the 'what' of closing material cycles. For instance, the reports from the Ellen MacArthur Foundation (e.g. EMF, 2013) present a framework where both technical and biological nutrients could be circularly managed. The examples of circular economy approaches provided in the existing literature also focus on either niche interventions or of large corporations closing material cycles within the value chains they control, thereby exacerbating existing inequalities in access to resources and capital (Schröder et al., 2018). According to the current CE discourse and practice, large corporations would be able to not only significantly influence the conceptualisation but also lead the transformation to the circular economy.

We do believe the transformation to a CE has to happen at scale and led by businesses (as envisaged in most of the existing literature). But it is not clear how businesses that gain from current linear models of make-use-throw would suddenly make the transformative leap? We also believe that such conceptualisation creates a limited understanding of the processes that contribute to closing material cycles and has the potential to lead to sub-optimal outcomes. For instance, due to the presence of the large informal sector in managing waste in most urban areas, it is critical to think about a framework that also includes these actors. Also, the roles of the government and the policymakers are largely ignored in the current framing. Furthermore, by focusing on the value chains of large businesses, there is limited attention to potential approaches at the level of a city, let alone a country.

In this chapter we have developed a conceptual framework that combines CE models and situates them in the variable power geometries of actors involved in these material flows to achieve a closing of material cycles. We then describe the narratives on waste management that exist at the supra-national, national and local levels. While characterising the national and local narratives, we focus on India and a large Indian city (Delhi), respectively. At the same time, we examine the impact of a focus on techno-managerial solutions on the vulnerable and marginal communities in urban areas. We believe that this chapter provides a new and much needed framework for analysing urban waste management, emphasises the need for such a framework, brings out its key components, and illustrates its relevance for India.

This chapter brings together two analytical shifts. The first shift is from a technical to political analysis: as most of the circular economy narrative is about materials, nutrients, metals etc. rather than people, our framework shows how people can be brought to the centre of CE. The second shift is from single to multi-scale analysis. The proposed framework shows that the narratives at the local, national and international levels interact with each other in several ways, on occasions reinforcing each other while on others contradicting each other. Such dynamics between the narratives have material consequences not only for the lives of people who are involved in waste management but also for the formulation of policies that are driven by dominant narratives. Our analysis suggests that the development of dominant narratives is a result of the contestations in the discursive arena, and as a result it is critical that the framework relies on a multi-scale analysis.

This chapter shows that building an economy which is both more circular and more equitable requires understanding the actors that play a role in these multiple circuits and understanding their priorities. It shows that these priorities are diverse and often in conflict with each other, particularly over the role of the informal sector, that there is a potential for building new alliances adopting less wasteful and more equal policies, and that identifying co-benefits is central for such an approach. In the following sections we describe and analyse the different actor constellations involved in closing material cycles. The focus on an actor-centric approach also allows us to develop and analyse several narratives and power relations that are associated with closing material cycles.

An actor-based circular economy framework

The transformation from a linear to a circular economy will require not only an environmental but also a social and economic restructuring of production and consumption patterns. Such systemic restructuring would involve numerous actors. Some of these actors are likely to benefit while others are likely to lose from such restructuring. As a result, some actors are likely to block while others are likely to drive the transformation to a CE. The actors who are likely to benefit from the transformation to a circular economy would drive for the transformation to occur in a manner in which they appropriate the gains from the transformation. Therefore, it is critical to develop an actor-centric framework for analysing the transformation from a linear to a circular system.

A starting point for the transformation to a CE is effective waste management. In most developed countries, policies towards closing material cycles have largely emanated as a reform of existing waste management policies. Policies aimed at closing material cycles are inextricably linked to the reformulation of the existing waste management policies. In India, the recently amended waste management rules are also based on the waste management hierarchy. However, distinct from most developed countries, a large part of the waste management in developing countries is handled by the informal sector. Therefore a critical uncertainty that emanates from the efforts at closing the material cycles is regarding the role of the informal sector in waste management (Wilson, Velis and Cheeseman, 2006). Does the transition to a CE marginalise the informal sector or does it use the strengths of the informal sector for cost-effective solutions? This question is at the heart of a just and equitable transformation as well as the focus of any actor-centric framework.

The informal sector, despite its informality, is well organised. Also, the services provided are efficient as well as convenient – two necessary characteristics of waste management systems globally. In addition to recycling and waste management service at low cost, the informal sector provides employment to urban poor in labour-intensive processes for collection, manual segregation and recycling (Agarwal et al., 2002). Despite the low-cost recycling and waste management service as well as the employment creation potential of the informal sector, it is discouraged considering the potentially negative health and environmental impact of unsafe recycling practices (Annamalai, 2015). Specifically, in the case of e-waste, environmentally sound recycling and enhanced resource recovery are the key reasons for establishing a formal waste management system in developing countries (Arora, Chaturvedi and Killguss, 2010:91). This creates a dichotomy in the broader narrative of recycling and closing the loop, with stakeholders divided between formal vs informal, high-tech vs low-tech and large private sector vs small or cooperatives of individual informal recyclers and waste managers.

Due to the complexity and dynamics involved in different circuits of the circular economy, each suggesting different pathways to different sustainabilities, it is argued that 'lock-in' to a powerful narrative and associated pathway can exclude others. Therefore, there is a need to 'open up' and make space for more plural and dynamic sustainabilities, challenging dominant narratives and pathways, and highlighting alternatives, including those reflecting the perspectives and priorities of poor and marginalised people (STEPS Briefing, undated). The STEPS 'pathways approach' developed by Leach, Scoones and Stirling (2010:ix) provides a useful analytical tool to understand the complex and dynamic narratives and identify the varied pathways of the CE. As noted by the authors (ibid:168) critical to evaluating alternative pathways is the need to emphasise those pathways which 'recognise and support the goals of people who are struggling to move out of poverty and marginalisation'. Applying this perspective to the circuits of CE, the informal enterprises engaged in repair, second-hand trade and recycling of waste gain importance in the scenario where they are compared to formal sector recyclers. As noted by Leach, Scoones and Stirling (2010:156) pathways exist in an environment of conditional plurality, i.e. the multiple approaches and solutions are dependent on the varied framings of problems. Thus, identification of the varied actors, their networks and objectives is crucial to understanding their specific framings. Leach, Scoones and Stirling (2010:65–66) note the dual effect of 'governance', which is defined by the authors as 'the intersection of power, politics and institutions', on both the framing of problems as well as the choice of solutions/pathways.

The barriers to informal sector pathways based on power dynamics and politics can be analysed using the power-cube framework developed by Gaventa (2006). The relative power and influence of various actors is analysed across spaces (closed, invited, created), levels (local, national, global) and different forms (visible, invisible and hidden) of power. Gaventa (ibid:26) notes that the analysis of spaces should seek to understand how the spaces were created, who created the spaces, and what the engagement terms are. Also, while each level – local, national, global – presents opportunities for participation, interconnections and linkages among the actors at different levels are essential for effective change. Batliwala (2002, cited in Gaventa,

2006:28) also notes the importance of 'vertical links between those organisations doing advocacy at an international level, often led or supported by international NGOs, with those working to build social movements or alternative strategies for change at the more local levels'. Similarly, the study of different forms of power aims to identify visible and invisible manifestations of power that influence the values and narratives of different actors. Leach, Scoones and Stirling (2010:169) also note that a pathways approach is not aimed to only assess the varied options, but also focuses on building pathways that are marginalised. This involves opening policy processes and spaces, including conceptual, bureaucratic, invited, popular and practical spaces (ibid:138). Additionally, different forms of citizen engagement and mobilisation ranging from protests, media and legal system engagement may be used to broaden pathways. Equally important for effective change is the development of alliances within and across different actors and levels, as highlighted by Gaventa (2006:28). This is also noted by Chaturvedi, Vijayalakshmi and Nijhawan (2015:28) as being crucial for sustaining pathways.

The varied actors involved in the different circuits of a circular economy can be divided into government, business and civil society or NGOs at local, national and supra-national levels. These actors have different levels of influence and can have aligned or conflicting interests. Table 2.1 presents Indian and international stakeholders in the repair, refurbish, recycle and incineration circuits:

TABLE 2.1 Actors involved in the multiple circular economy circuits.

	Business	*Government*	*NGOs*
Repair and reuse circuit	Small businesses, traders of second hand products	Local government mostly	Local and national level NGOs few international NGOs like GAIA
Refurbishing circuit	Private medium-sized companies	Local and national governments	Local and national NGOs promoting refurbishing
Formal recycling circuit	Large corporations with agreement with formal recyclers	Local, national and supra-national governments like UN, European Union.	NGOs like World Economic Forum, EMF etc.
Informal recycling circuit	Small informal sector recyclers	Local governments and national governments	NGOs at local and national level in addition to few international NGOs like GAIA
Incineration circuit	Waste management companies charging for waste incinerated to generate power or reduce volume of waste	Local, national and supra-national governments like UN, European Union.	NGOs like World Economic Forum, EMF etc.

Multiple circuits – multiple narratives

It is clear from the previous discussion that while stakeholder groups have individual interests and agendas, there is active alliance building for promoting a common narrative to influence policy and implementation. The alliances can range from local level to supra-national level. For instance, the World Economic Forum representing multinational firms (businesses) is working with the United Nations, the European Union (governments) and the Ellen MacArthur Foundation (NGOs) to promote the idea of a private sector, technology-led circular economy approach (one of the many circuits of the circular economy) influencing national and local level government. Another approach to the circular economy involving the informal sector for local level repair and recycling (another circuit of the circular economy) has the alliance of local, national Indian NGOs like Chintan, Centre for Science and Environment (CSE) and international NGOs like Global Alliance for Incinerator Alternatives (GAIA).

It is evident that there is a need to better understand the dynamics within the circuits as well as how they interact with the policy landscape. The conceptual framework presented in Figure 2.1 highlights how the CE has local, national and supra-national level dimensions through the flow of raw materials, products and waste generation within a country and between countries through exports and imports.

The framework suggests that due to the flows of materials between cities and nations, closing material cycles cannot happen within local or national boundaries. The flows of materials across local and national boundaries implies that, in a material sense, cities as well as nations are open systems (Chaturvedi and McMurray, 2015). At the same time, actors from different sectors and the state operating at different levels have a constant interaction between the narratives espoused by them. All actors have a certain set of objectives and priorities as well as a conception of the transformation to a CE. This interaction between the actors in the material and discursive arena leads to the development of alliances between the actors who coalesce around shared conceptions of the future as well as shared objectives.

Different narratives have emerged at different scales on the transformation to a circular economy. The narratives reflect the conception of economic, social and political leanings along with priorities and interests of the actors. We classify these narratives according to scale because of the scope of influence of the actors who espouse the narratives. While this classification might seem occasionally restricted as these actors occupy and influence several scales simultaneously, the classification according to scale allows for analytical tractability. Also, in most cases the restriction is also real as it reflects the priorities of the actors and their sphere of influence.

Supra-national narratives

The narratives at the supra-national level are largely based on the premise that the circular economy is critically linked to the sustainability of the planet. For instance,

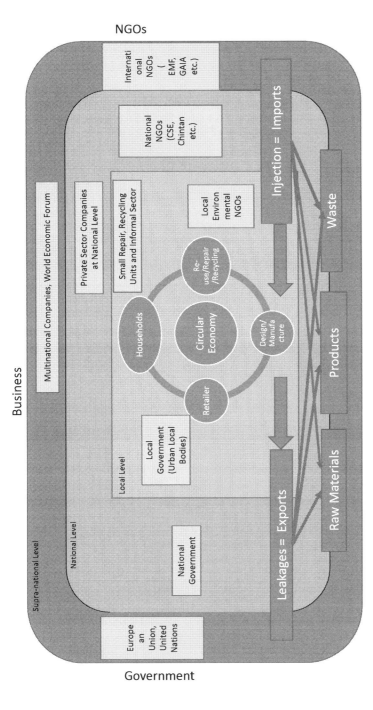

FIGURE 2.1 Actor-based circular economy conceptual framework.

the Ellen MacArthur Foundation (EMF) states that the CE is restorative and regenerative by design, and which aims to keep products, components and materials at their highest utility and value always, distinguishing between technical and biological cycles. The EMF's Circular Economy 100 is an innovation programme established to enable organisations to develop new opportunities and realise their circular economy ambit, which is claimed to be a €1.8 trillion circular economy opportunity. EMF has published several studies focusing on the role of mostly large private sector organisations in achieving circular economy objectives building on technology and private capital (EMF, 2017).

At the same time, these narratives build on notions of resource efficiency, resource security and environmental considerations. For instance, the European Union focuses on the resource efficiency and resource security agenda as the cornerstone of the transformation to a circular economy (EEA, 2016). Of course, these considerations also have co-benefits for the economy and society, but those are not centre stage. The co-benefits are a useful political device for developing alliances with actors who might not be explicitly linked to the environmental agenda. The key aspects of the supra-national narratives are the following:

1. Businesses, mostly large corporations, have the potential to lead the transformation to a circular economy at scale.
2. Innovations, mostly technological and material and design specific but also related to business models, such as product service systems (PSS), are critical for the transformation to a CE.
3. Consumers, mostly passive actors who respond to marketing, advertising and the incentives that are embedded in business models and react in accordance to these incentives.
4. There is limited focus on social and political relations of the different actors who are involved in the large-scale transformations. Technology and innovations systems are not embedded in the social and political realities in which the corporations operate. Also, there is limited focus on questioning the fundamental drivers that have led to the 'throw-away' economy and how these would be circumvented by the same actors. It is almost assumed that the enlightened self-interest of the corporation would lead to the transformation to a CE.
5. As a result of these factors, there are limited efforts at questioning the basic economic model which has led to the creation of a 'throw-away' economy.

Of course, competing narratives question these models based on environmental concerns, either based on the assumptions or on the belief that large corporations can be at the centre stage of the transformation to a circular economy. Several assumptions are also questioned based on the lived realities that emerge from the experience of developing countries, where waste management is deeply political and the models for closing material cycles are determined to a large extent by social and economic considerations. At the supra-national level, GAIA is a worldwide alliance of more than 800 grassroots groups, non-governmental organisations and individuals in over

90 countries, campaigning against the establishment of incinerators in developing countries. However, the narratives against technology and large private sector-led environmental solutions like the waste-to-energy are not influential enough at a supra-national level. For instance, despite several years of legal battles, GAIA failed to stop construction and continued operation of existing waste-to-energy plants in India based on incineration technology (GAIA, 2017).

We describe some of the national and sub-national level narratives based on our work in India which highlight the possibility of preference for one CE approach over another, leading to a potentially negative impact on the stakeholders involved in the other approach to CE.

National narratives

In India, a large part of the focus on transformation to closed material cycles is inspired by the inability to manage waste effectively. The key campaign 'Swachh Bharat' of the current government focuses on cleaning India. While there has been a clear failure in effective waste management (Joshi et al., 2016), the narratives of zero waste, waste to wealth and waste to energy are extremely powerful and widespread with significant private sector investor interest following lucrative policies supporting waste-to-energy through feed-in tariffs (Prasad, 2016). By declaring waste as a repository of value, it is wished that waste would eliminate itself through benefactors interested in the embedded materials and energy. A large part of the focus on waste-to-wealth and waste-to-energy is also driven by the belief that it would draw in the private sector to solve the problems that have thus far been ineffectively dealt by the public sector. Public-private partnerships are therefore seen as the panacea for managing the waste crisis. The key aspects of the national narrative in India are the following:

1. Large corporations are going to lead the transformation regarding the waste management problem and could provide the technologies that enable recovery of resources and energy.
2. While there are significant overlaps in the supra-national and national narratives, there is still no large-scale policy reform that puts resource security and the CE at the centre stage of industrial policy.
3. The public sector should be involved in creating the right policy environment for the management of waste and provide the incentives to the private sector to perform efficiently and effectively.
4. The consumer has been given the moral responsibility by equating cleanliness with godliness. There is limited attention on how consumer behaviour would change without an adequate infrastructure that enables the right behaviour or without even thinking about the convenience of recycling.
5. The contestations for materials is acknowledged in the legal framework by invoking the informal sector and appeals at mainstreaming informality without making any efforts regarding the requisite institutional changes.

Like the international narratives, the national narratives face significant challenges from actors who are involved in shaping policies as well as involved in waste management work in the cities. While there are concerted efforts to generate evidence from international and national experiences about the limitations of waste as a source of energy, the imaginaries of a waste-derived energy infrastructure are still pursued with much vigour in policy circles. Even if the energy generation potential of the entire Indian MSW, estimated at around 511 MW (Planning Commission, 2014) would be achieved, it would meet less than 10 per cent of Delhi's power demand, which reaches 6,044 MW during the summer season (PTI, 2016). This evidence belies the focus on waste-to-energy as a potential contributor to meeting the energy demand of large cities, which has been the basis of media campaigns of several stakeholders. The most critical counter-narratives at the national level in India are all rooted in the environmental challenges of effectively governing the waste-to-energy infrastructure and questioning the technical appropriateness of such technologies in India.

Local urban narratives

At the local city level, the prime objective of the citizens as well as the urban local bodies is a clean and hygienic environment. There is limited interest in closing material cycles because cities are essentially open systems in a material sense – products enter and leave the boundaries of the city. Closing material cycles for a city government, as a result, is quite challenging. However, there are significant efforts at closing material cycles *within* city boundaries, especially in the large metropolitan areas because this activity creates jobs for a significant part of the working population. For instance, in Bangalore, several material recovery facilities have been set up to channelise the recyclables for effective recovery of embedded resources. In large urban areas, waste management provides a barrier-free entry to the job market and entrepreneurship.

The dominant narrative, however, is not related to jobs but rather concerns clean and hygienic cities. For instance, politicians invoke the image of London and Paris as examples of aspirational imageries for city managers. Also, there is a clear focus on involving the private sector – in this case large waste management companies – to provide solutions for waste management. There is only limited evidence for the strong belief in the capacities of the private sector to deliver an impact, especially in the Indian context.

Due to this tension between privatisation efforts as well as the job-creating potential of waste management (largely in the informal sector), there are quite strong conflicts in both the material and discursive arenas. The key aspects of the narratives at the city level are:

1. The private sector, especially waste management companies, can solve the problem at scale in large cities.
2. Technological solutions are likely to play a critical role in achieving progress.

3. At the same time, equally dominant are the voices of informal actors as well as environmental NGOs that question the environmental performance and appropriateness of technological choices.
4. At the city level, there is limited interest amongst the city managers on grand narratives of closing material cycles – they are driven more by the national imaginaries of zero-waste cities, waste-to-wealth as well as waste-to-energy.

The recent developments in the waste management sector in India are creating uncertainty for the informal sector, while formal large-scale companies are being promoted to introduce technology-dominated solutions for waste management and recycling (Gidwani and Reddy, 2011). It is observed that only a few local governments have engaged and worked closely with informal sector actors, while most prefer to work in favour of large private companies (Cavé, 2012).

While the technology and private sector-led approach to circular economy dominates, there are emerging models of organising waste pickers, local communities and activist groups that can provide waste management and recycling without compromising on environmental quality and health. Decentralised small-scale, informal and sector-led recycling with the support of low-cost solutions provide socially inclusive alternatives to high-cost market-led solutions (Demaria and Schindler, 2016).

The dynamics of narratives

How do certain narratives emerge as the dominant narrative? Our framework allows for a response to such queries by arguing that actor-alliances that converge on certain aspects of closing material cycles enable the development of dominant narratives. Such convergence can happen within a level, or the alliances can be forged amongst actors at different levels. For instance, the narrative that the private sector can deliver effective waste management solutions largely emerges from the alliance between local governments, formal private sector actors at the local, national and supra-national level, and the policies of the central governments that are rooted in the belief of the ability of private sector to overcome the challenges of local governments (see Chaturvedi, Arora and Saluja, 2015). Similarly, the contributions and primacy of the informal sector in waste management in large urban cities are driven by the alliance of civil society organisations, which are bilateral and multilateral agencies active at several levels. We classify these alliances as intra-level alliances as well as inter-level alliances. The intra-level alliances are formed when actors with divergent objectives coalesce around specific aspects of a circular economy. For instance, at the local level such alliances can be forged between the local government and the private waste management companies. A competing alliance can be forged between the civil society organisations as well as grass roots movements of the informal sector that advocate for local decentralised solutions using the resourcefulness of the informal sector. At the local level, these contestations could lead to material conflict between the formal and informal sectors, a case which is evident in Delhi. At the national and supra-national levels, similar alliances can be formed to influence the national policy processes.

Existing literature and the experience of policy-making processes in India clearly suggest that there are significant interactions between the actors that operate at different levels. In certain cases, the same actors operate at different levels, making the analysis of narratives and dynamics quite complex. To put order into this complexity, it is critical to analyse the potential alliances that can emerge between actors who are operating at different levels as distinct from intra-level alliances. For instance, the national government in India and governments of other developing economies are influenced by the best practices abroad when drawing up waste management policies. In some cases, however, the narratives at the supra-national level heavily influence discussions on policy options at the national level and impact the implementation of the policies at the local level. This influence is expected, but could be used to explain the choice of policies and implementation strategies that might lead to sub-optimal outcomes locally. As mentioned previously, the focus on waste-to-energy as a potential win-win situation for waste management as well as solving the energy crisis is a case in point.

While there are possibilities for certain narratives to gain strength due to the forging of intra- and inter-level alliances, as a corollary, it is also clear that certain narratives fade away due to the inability of their proponents to forge such alliances. A clear case in point is the role of the informal sector. Despite the overwhelming evidence that the informal sector is extremely resourceful and contributes significantly to managing waste as well as in closing material cycles, its role is not at the centre stage of any policy directive at the national level in India. This is evident from the new Solid Waste Management Rules announced in 2016: although the informal sector is mentioned as a critical actor and there is an acknowledgement of its contributions, the focus of implementation and the incentives are designed in a manner that it is further marginalised.

In certain cases, although the focus of the narrative might be on the same aspect of a CE – e.g. closing material cycles – there might be conflicts for the actors who have the potential to drive the change. For instance, at the international level, the discourse on closing material cycles is heavily influenced by the experiences of the now developed economies. As a result, the emerging narratives focus on the primacy of the large corporations and private sector entities involved in mining, production, consumption and waste management. However, at the local level in developing countries, it is the informal sector that is involved not only in waste management, but also in repair and refurbishment activities that have significant positive impacts on closing material flows. Our analytical framework suggests that the role of the large corporations is likely to gain traction because of its ability to gain the support of the actors who are operating with similar objectives at the local, national and supranational levels, while the role of the informal sector is likely to fade away as the support at the supra-national level might be missing. Table 2.2 summarises some of the narratives that have emerged.

TABLE 2.2 Dynamics of narratives.

Level	Synergies across levels	Conflicts across levels
Supra-national National Local	Large formal companies as drivers for closing material cycles Technology as a driver for the transformation to a circular economy Failure of state-supported models	Inability of informal sector models in closing material cycles Informal sector practices are environmentally unsafe and do not comply with labour standards Discrediting of decentralised models based on the innovations in the informal sector

Power analysis

This section evaluates the reasons for the dominance of the market-led narrative that in some cases is leading to the exclusion/marginalisation of the informal sector's narrative in developing countries. While some visible reasons for this are well documented in the literature (Ezeah, Fazakerley and Roberts, 2013) and include the environmental, safety and health implications of the informal sector, there is a need to explore other, often hidden, socio-political causes. As mentioned previously, a review of the choice of circular economy narratives necessitates an understanding of not only the actors, discourse and narratives but also the power dynamics (Leach, Scoones and Stirling, 2010:129), which are analysed across different spaces, levels and forms of power (Gaventa, 2006). We argue that despite the emergence of new spaces and opportunities for engaging stakeholders in policy processes, simply creating new institutional arrangements will not necessarily result in greater inclusion or pro-poor policy change. Rather, much will depend on the nature of the power relations that surround and imbue these new, potentially more democratic, spaces.

The first aspect of the power-cube framework seeks to review spaces as closed, invited or created. These spaces can be evaluated at the local, national and global levels. At the national level, the policy process for the drafting of e-waste rules has become a multi-stakeholder based model for agenda setting, policy development and discussion. This Indian national space can be considered as a created space and was formed by actors including international, national and local NGOs, industry associations and international agencies like the UN and Deutsche Gesellschaft für Internationale Zusammenarbeit (GIZ). While the informal recycling sector was engaged in dialogue and consultations, the other key group of reuse and repair enterprises was not adequately represented, and the formal recycling sector gained legitimacy while the informal sector was influenced to engage with the formal sector. Although, the national policy space is a created space from the viewpoint of NGOs, international organisations and the private sector, it is an invited space with regards to the varied organisations engaged in the informal sector, and the greater focus on formal recycling has led to a lack of acknowledgement of the informal repair, reuse and recycling sector.

At the global level, formal and private waste management companies dominate the narratives related to the environmentally sound management of e-waste. For instance, a 2007 joint statement by a group of international companies and NGOs (Greenpeace, 2007) provides an example of the global dynamics wherein financial incentives/market mechanisms are crucial for improvement in formal e-waste management. Some international informal sector initiatives like Global Alliance of Waste Pickers and WIEGO have facilitated recognition of waste collector cooperatives as well as the creation of cooperatively owned material recycling facilities (GlobalRec.org, 2014), yet similar mechanisms for the informal repair and reuse sector are not well developed and are limited for e-waste recyclers.

In the analysis of power at the local level, it is important to recognise both the policy development and implementing agencies. While the policy development process is influenced by the actors at local, national and international levels, the implementation necessarily happens at the local level. As a result, at the local level, the actors influential at the material and discursive arenas converge. As documented elsewhere, this creates opportunities for conflict as well as collaboration between the two different pathways – led by the market and the informal sector (Arora et al., 2010). While national level policies, including the National Environment Policy of 2006, recognise the need for legal recognition of the informal recycling sector in addition to strengthening their access to finance and technology (WIEGO, n.d.), at the local level varied levels of engagement are visible. However, these opportunities for conflict and cooperation are significantly influenced by the actors with significant convening power at the local level – the State Pollution Control Boards responsible for the implementation of the law as well as the urban local bodies responsible for waste management in the city. Hence, the commitment of the implementing agency at the local level influences the engagement of various stakeholders (Toxics Link, 2014:30–47), making access and representation of the informal sector to important decision-making bodies at the local level dependent on the priorities of the local implementing agencies.

It is also essential to note the invisible and hidden forms of power that influence the visible discourse and narratives. The presence of ideologies, values and perceptions around waste and the informal sector can be examined through the widely reported and circulated images of the detrimental effects of informal recycling, which limit a deeper evaluation of the varied aspects and skills present within the informal sector, including repair and reuse. Combined with the ideas of a centralised and hi-tech space for the processing of e-waste, the rudimentary and small shops for repair, reuse or recycle receive little emphasis or credit for the key role they are playing in e-waste management in developing countries. This variation in the physical portrayal of the informal sector also influences the circular economy discourse towards more centralised and formal approaches to e-waste. This is noted by Chaturvedi, Vijayalakshmi and Nijhawan (2015:12) as a discrediting of the informal sector, involving 'branding the informal sector as drug addicts, thieves and relics of a pre-modern society who should not be allowed space in a modern world class city'. Additionally, for the repair and reuse sector, illegality is a key barrier with the widespread market for counterfeit goods and parts in India. The counterfeit electronics market in India is growing at twice the rate of the genuine product market, with Delhi city contributing almost 75 per cent of the trade in counterfeit products (Vikram, 2013). Thus, distinguishing between the legal and useful operations versus the illegal aspects of informal enterprises also hinders greater recognition by and support from local and international actors.

Yet, despite these barriers the informal sector continues to exist, which highlights the contrasting dynamics at the local and global levels. At the local level, the widespread markets for repair and reuse continue to thrive in a challenging environment. Their local material power through access to e-waste, however, has limited influence on the discourse and dominant framings of the circular economy at the national and global levels. These socio-political barriers, hence, inhibit greater inclusivity and opportunities for opening the debate on the informal sector's role in the circular economy.

Conclusion

In this chapter, we argue that the focus of closing material cycles must move away from a big business-led approach if societies want to transform from a linear to a circular economy. Our proposition is based on the premise that such a transformation is essentially a political process. The focus on an actor-led model would have to be evaluated for its political feasibility. As a result, to develop an analytical lens, we demonstrate that it is critical to identify the many narratives that emerge from different stakeholders who are involved in this transformative process. We therefore conclude that to analytically assess the possibilities of a transformation to a CE, it is critical to take an actor-centric approach. We also contend that the actors involved in the transformation to a CE are spatially distributed within a city and within a nation state as well as beyond the boundaries of a nation state. As a result, we make the case for analysing narratives at different scales to pin down potential actor constellations that would drive the multiple conceptions of a CE. It is clear from the previous discussion that the dominant narratives are likely to be those that bundle interests across scales.

Our analysis is suggestive of a complex relationship between the material and discursive arenas and that CE narratives are currently in flux. As a result, it is critical to focus on the politics of the CE. If evidence-based research has to inform conceptions of a CE, now is probably the most important juncture to actively pursue the agenda, before the alliances are stabilised and certain narratives emerge as the clear front-runners.

We believe that the unique contribution of this research is two-fold. First, our analytical approach combines for the first time in the literature on circular economy a multi-actor, multi-scale model. This approach allows us to engage with multiple narratives at the local, national and supra-national levels. Second, our results suggest that contestations in the material and discursive arena of waste management must be understood beyond the boundaries of an urban agglomeration. The waste management narratives in a large city in India, or for that matter in the UK, are shaped by actors not only within the city boundaries but also in other parts of the world. As a result, solutions that get prioritised must also be evaluated with an analytical lens that incorporates these different influencers. We believe that our approach allows for an explanation of some of the innocuous choices of solutions, for example the focus on the incineration of waste, made by policymakers regarding waste management in India and elsewhere.

Note

1 An extended version of this chapter was published as STEPS Working Paper 94 Available at: https://steps-centre.org/publication/many-circuits-circular-economy/

References

Agarwal, R., Gupta, S.K., Sarkar, P. and Ayushman (2002) *Recycling Responsibility: traditional systems and new challenges of urban solid waste in India*. New Delhi: Toxics Link. [Online] Available at: http://toxicslink.org/docs/munispalwaste/Recycling%20Responsibility_mail.pdf [Accessed 21 February 2017].

Annamalai, J. (2015) Occupational health hazards related to informal recycling of e-waste in India: an overview. *Indian Journal of Occupational and Environmental*, 19 (1), 61–65.

AroraR., Chaturvedi, A. and Killguss, U. (2010) Environmentally sound e-waste recycling in India: mainstreaming the informal sector in the formal recycling system. *Regional Development Dialogue*, 31 (2), 90–100.

Cavé, J. (2012) Urban solid waste in southern countries: from a blurred object to common pool resources. World ISWA Congress, September 2012, Florence, Italy. [Online] Available at: https://halshs.archivesouvertes.fr/hal-00737461/document

Chaturvedi, A. and McMurray, N. (2015) *China's Emergence as a Global Recycling Hub: what does it mean for circular economy approaches elsewhere?* IDS Evidence Report 146. Brighton: Institute of Development Studies. [Online] Available at: https://opendocs.ids.ac.uk/opendocs/handle/123456789/6932

Chaturvedi, A., Arora, R. and Ahmed, S. (2010) Policy cycle – evolution of e-waste management and handling rules. In: *National Conference on Sustainable Management of E-Waste*. New Delhi: GTZ-ASEM, pp. 3–4. [Online] Available at: www.weeerecycle.in/publications/research_papers/Policy_Cycle-EWaste_final_10_12_06.pdf

Chaturvedi, A., Vijayalakshmi, K. and Nijhawan, S. (2015) *Scenarios of Waste and Resource Management: for cities in India and elsewhere*. Institute of Development Studies. IDS Evidence Report 114. [Online] Available at: www.ids.ac.uk/publication/scenarios-of-waste-and-resource-management-for-cities-in-india-and-elsewhere [Accessed 27 May 2015].

Chaturvedi, A., Arora, R. and Saluja, M. (2015) Private sector and waste management in Delhi: a political economy perspective. *IDS Bulletin*, 46 (3), 7–15.

Demaria, F. and Schindler, S. (2016) Contesting urban metabolism: struggles over waste-to-energy in Delhi, India. *Antipode*, 48, 293–313.

European Environment Agency (EEA) (2016) *More from Less – Material Resource Efficiency in Europe: overview of policies, instruments and targets in 32 countries.* [Online] Available at: https://www.eea.europa.eu/publications/more-from-less

EMF (2013) *Towards a Circular Economy: economic and business rationale for an accelerated transition.* Cowes, Isle of Wight: Ellen MacArthur Foundation.

EzeahC., Fazakerley, J.A. and Roberts, C.L. (2013) Emerging trends in informal sector recycling in developing and transition countries. *Waste Management*, 33 (2013), 2509–2519.

Gaventa, J. (2005) Reflections on the uses of the 'Power Cube' approach for analyzing the spaces, places and dynamics of civil society participation and engagement. [Online] Available at: www.participatorymethods.org/sites/participatorymethods.org/files/reflections_on_uses_powercube.pdf

Gidwani, V. and Reddy, R.N. (2011) The afterlives of 'waste': notes from India for a minor history of capitalist surplus. *Antipode, 43* (5), 1625–1658.

Global Alliance for Incinerator Alternatives (GAIA) (2017) NGT's judgment on Okhla Incinerator Plant case: a judicial overreach! [Online] Available at: www.no-burn.org/why-the-ngt-okhla-order-sets-a-bad-precedent-for-indian-environmental-jurisprudence/

GlobalRec.org (2014) *Four Pilot Projects in the Works: an update from the South African Waste Pickers' Association.* [Online] Available at: http://globalrec.org/2014/06/26/four-key-pilot-projects-operating-under-sawpa/

Greenpeace (2007) Joint statement by a group of industry and NGOs on producer responsibility for waste electrical and electronic equipment. Greenpeace. [Online] Available at: www.greenpeace.org/international/Global/international/planet-2/report/2007/3/joint-statement-by-a-group-of.pdf

Imran, M., Haydar, S., Kim, J., Awan, M. and Bhatti, A. (2017) E-waste flows, resource recovery and improvement of legal framework in Pakistan. *Resources, Conservation and Recycling*, 125, 131–138.

Joshi, R., Ahmed, S. and Ng, C.A. (2016) Status and challenges of municipal solid waste management in India: a review. *Cogent Environmental Science*, 2 (1), 1139434. [Online] Available at: http://home.iitk.ac.in/~anubha/H13.pdfdoi: 10.1080/23311843.2016.1139434

KumarS., Smith, S.R., Fowler, G., Velis, C., Kumar, S.J., Arya, S., Kumar, R. and Cheeseman, C. (2017) Challenges and opportunities associated with waste management in India. *Royal Society Open Science, 4* (3), 160764. [Online] Available at: https://www.ncbi.nlm.nih.gov/pmc/articles/PMC5383819/ [Accessed 26 September 2018].

Leach, M., Scoones, I. and Stirling, A. (2010) *Dynamic Sustainabilities.* London: Earthscan

Planning Commission (2014) *Report of the Task Force on Waste to Energy.* Volume 1. New Delhi, India. [Online] Available at: http://planningcommission.nic.in/reports/genrep/rep_wte1205.pdf [Accessed 21 February 2017].

Prasad, G.P. (2016) Waste-to-energy projects see revival in investor interest. *Livemint*, 23 March, 2016. [Online] Available at: www.livemint.com/Industry/B9q700vtN6YL5jxndS3rjL/Wastetoenergy-projects-see-revival-in-investor-interest.html.

PTI (2016) Delhi's power demand rises to a new record of 6044 MW. *Financial Express*, 19 May 2016. [Online] Available at: www.financialexpress.com/economy/delhis-power-demand-rises-to-a-new-record-of-6044-mw/259844/.

Schröder, P., Dewick, P., Kusi-Sarpong, S. and Hofstetter, J.S. (2018) Circular economy and power relations in global value chains: tensions and trade-offs for lower income countries. *Resources, Conservation and Recycling*, 136, 77–78.

STEPS Briefing (2011) *The Pathways Approach of the Centre.* [Online] Available at: http://steps-centre.org/wp-content/uploads/STEPS_Pathways_online1.pdf

Toxics Link (2014) *Time to Reboot.* Delhi: Toxics Link. [Online] Available at: http://toxicslink.org/docs/Time-to-Reboot.pdf

Vikram, K. (2013) Industry report reveals Delhi contributes 75 per cent to India's booming market in counterfeit goods. [Online] Available at: www.dailymail.co.uk/indiahome/india news/article-2342341/Black-market-Capital-Industry-report-reveals-Delhi-con tributes-75-cent-Indias-Rs-45-000-counterfeit-goods-trade.html [Accessed 22 July 2015].

WEF (2017) *Shaping the Future of Environment and Natural Resource Security*. [Online] Available at: www.weforum.org/system-initiatives/environment-and-natural-resource-security

Wilson, D. C., Velis, C. and Cheeseman, C.R. (2006) Role of informal sector recycling in waste management in developing countries. *Habitat International*, 30, 797–808.

World Economic Forum (2014) *Towards the Circular Economy: accelerating the scale-up across global supply chains*. Geneva: World Economic Forum. [Online] Available at: www.weforum.org/docs/WEF_ENV_TowardsCircularEconomy_Report_2014.pdf [Accessed 21 February 2017].

3

THE POLITICS OF MARINE PLASTICS POLLUTION

Patrick Schröder and Victoria Chillcott

Introduction: out of control

Plastics pollution of rivers and the marine environment has been identified as an emerging Anthropocene risk and planetary boundary threat (Villarrubia-Gómez, Cornell and Fabres, 2017). Estimates are that during 2010, up to 12.7 million tonnes (Mt) of mismanaged land-based plastic waste entered the oceans (Jambeck et al., 2015). Microplastics and their impact on marine life (Carson et al., 2011) and on food chains highlight the links between ocean health and human health. Not only coastal areas, but also rivers are impacted and act as pathways of plastics pollution to the oceans (Schmidt, Krauth and Wagner, 2017). Plastics production is expected to double over the next 20 years. If current production and waste mismanagement trends continue, by 2050 it is estimated that roughly 12,000 Mt of plastic waste will be in landfills, the natural environment and marine systems (Geyer, Jambeck and Law, 2017). These figures show that human society collectively has lost control over plastics – undermining the sustainability of continued prosperity. From an economic perspective, plastic waste, in particular packaging, generates substantial negative externalities – the total natural capital cost to marine ecosystems of plastic littering has been conservatively valued at US$13 billion per year (UNEP, 2014) and this figure is expected to increase with strong volume growth in plastic production. Currently only about 14 per cent of all plastic packaging is collected for recycling after use, which results in a direct economic loss of US$80–120 billion per year (Ellen MacArthur Foundation, 2017). Recent data show that national plastic recycling rates in some countries are even in decline: in the US domestic plastic recycling fell from 9.1 per cent 2015 to 4.4 per cent in 2018 and could fall to a mere 2.9 per cent in 2019 (Dell, 2018).

To date, scientific research has focused on estimating the magnitude of the problem and understanding the potential impacts on marine organisms, food chains and human health. Response strategies, including optimism about the potential of circular economy approaches, are hindered by major gaps in scientific knowledge about the critical leakages, the systemic behaviour of the impacts, proximity and time scales of vulnerabilities and tipping points, and realistic ways to address the problem of plastic debris (Mendenhall, 2018). Marine environment experts, policymakers, and industry actors alike acknowledge the urgent need for interdisciplinary and transdisciplinary research (GESAMP, 2016) in order to better understand and act on the issue of plastic pollution.

At the international level, there have been several conventions which have tried addressing the issue of marine debris in the past, including the International Convention for the Prevention of Pollution from Ships (MARPOL) and the United Nations Convention on the Law of the Sea (UNCLOS), none of which have been able to curb the unending flow of plastic waste into our oceans (Leous and Parry, 2005). The oceans were formally discussed during the 2012 Rio+20 UN Conference on Sustainable Development, after previously being established as 'the common heritage of mankind' (Pardo, 1984 cited in Silver et al., 2015:136) and as a 'significant ecological frontier' (Steinberg, 2008 cited in ibid.). More recent global agreements concerned with marine plastic pollution include the Sustainable Development Goal 14 and its target 14.1, which seeks, by 2025, to 'prevent and significantly reduce marine pollution of all kinds, in particular from land-based activities, including marine debris and nutrient pollution'. Furthermore, a new proposal to amend the Basel Convention, which governs the international movement of waste materials, would reclassify scrap plastic under the category of 'wastes requiring special consideration', which would restrict international shipments of plastic waste (Staub, 2018).

Despite the SDGs' worthy goals and the tightening of the Basel Convention rules, it is unlikely that this framework be sufficient to solve the marine plastic challenge. The main reason is that complex national politics and vested corporate interests stand in the way of increased international collaboration and global coordination, which is vital to the efficient and effective mitigation of marine plastic pollution. In the Global North, it is mainly a lack of political will and no effective coordination between countries, in addition to fragmented environmental policies and corporate lobbying power of 'Big Plastic' and global brands, which has stymied attempts to stem the tide of plastics entering the seas. In the Global South, scarce resources as well as lack of expertise, manpower and technology prevent many possibilities to stop leakages of plastic into the environment. Lack of enforcement of existing waste management regulations is yet another issue, one that is especially relevant for developing country contexts. In the past, policy has mainly focused on near-coastal areas, while there has been little effort to address inland littering, open dumping and ineffective solid waste management policies that lead to marine debris (Leous and Parry, 2005). Overall, policy has only affected a minimal proportion of plastic pollution. There have been some gains from bottom-up governance, but these 'are not coming close to keeping pace with the rising environmental costs from the globalization of plastic' (Dauvergne, 2018:29).

Recent developments show there are mechanisms to affect policy, not necessarily through a linear transition from scientific evidence to evidence-based policymaking, but through the creation of transformative alliances which include NGOs, media, environmental activists, scientists and concerned citizens. The alliances and numerous activities which have created a feeling of collective public guilt about the state of the oceans, eventually, have led to some action from politicians, who are responding to the pressure from their constituents. How this occurred in the UK context, we analyse and explain in the following section.

What influenced the UK to tackle plastic pollution?

The UK Government announced in January 2018 they are working towards tackling plastic pollution in their 25-Year Environment Plan. Different actors have shown to have influenced this political action. A hybrid of the political climate, in particular around Brexit and public pressure, which increased after the *Blue Planet II* documentary series, can be seen to have affected the discourse around plastic pollution. Discourse around plastic pollution has increased in parliament and the media (see Figures 3.1 and 3.2) and it is necessary to understand what influence caused this change in political discourse on an issue of global importance.

What is interesting then is why the UK public became suddenly passionate about the issue of plastics, and what caused the shift in focus to put pressure on the government? The 'Blue Planet Effect', referring to the popular *Blue Planet II* documentary, which was frequently mentioned in the media, interviews, and in parliament debates (Great Britain, 2018) has been a main event. Policymakers, including the Conservative MP and Minister of State at the Department for Business, Energy and Industrial Strategy, remarked on the immense importance of the 2017 BBC documentary. *Blue Planet II* clearly demonstrated the effect that plastic pollution was having on the oceans and conjured conversation and debate about plastic pollution (Mail Online, 2018). The Secretary of State for Environment, Food and Rural Affairs was quoted as being 'haunted' by images from the series (Rawlinson, 2017). Following *Blue Planet II* in the winter of 2017, the media attention more than doubled (see Figure 3.2), including in

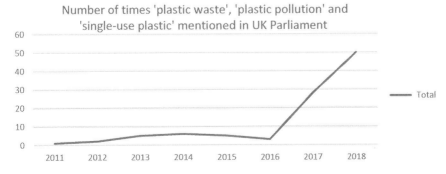

FIGURE 3.1 Line graph demonstrating the use of the terms relating to plastic referenced in the UK parliament.
Source: Chillcott, 2018.

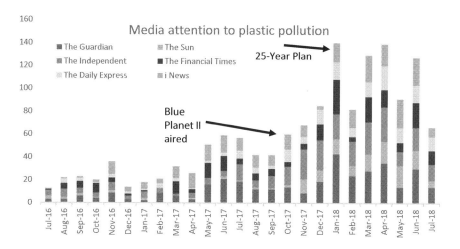

FIGURE 3.2 A bar chart demonstrating the increase in media attention around plastic pollution. Online articles counted. Note the references to the *Blue Planet II* programme and the UK government's 25-Year Environment Plan.
Source: Chillcott, 2018.

the Sun, and the Telegraph. This went beyond the usual papers like the Guardian and Independent, and thus incorporated different, more conservative readers.

Furthermore, due to the messy political climate following the Brexit referendum in June 2016, it has been insinuated that the government used this issue to be seen to be doing good (Chillcott, 2018). Brexit had daily news coverage with online media and newspapers dedicating entire sections to the topic (*The Guardian*, 2018; *The Sun*, 2018; *Evening Standard*, 2018). A Conservative MP remarked 'online searches around plastic recycling increased by 55 per cent' after *Blue Planet II* aired, and 'the government has been very keen to harness this enthusiasm, which has resulted in a number of exciting new policy initiatives to tackle this issue and create a long-term plan for our environment' (Chillcott, 2018). It is not clear whether it is a coincidence that, at the same time that Brexit was being substantially covered in the media, the *Blue Planet II* programme made the public aware of the plastic pollution problem, which led to the government tackling something that the public care about (Chillcott, 2018). Although unrelated to the issue, the political climate around Brexit arguably had an influence on the 25-Year Environment Plan.

Blue Planet II 'really helped establish that pressure [from] the constituencies that the current government really care about' (Greenpeace, 2018). In turn, the UK Government has called for suggestions on what it should do (Treasury, 2018a); therefore, the *Blue Planet II*'s impact on the public can be seen to have ultimately led to policy action. An action in the 25-Year Environment Plan was the launch of 'a call for evidence in 2018 seeking views on how the tax system or charges could reduce the amount of single use plastics waste' (Defra, 2018). The Treasury published the consultation 'Tackling the plastic pollution: using the tax system or charges to address single-use plastic waste' in March 2018 (Treasury, 2018a). This call for evidence was in collaboration with the 25-year plan and was responded to by more than 162,000 people, which included individuals in addition to campaign groups and companies. 'An unprecedented number of people have backed tough action against plastic waste' (Elgot, 2018).

In August 2018, the summary of responses was published, which included the proposed 'next steps' outlined in the Treasury's summary (Treasury, 2018b:30). The responses for the call of evidence to plastic pollution include:

- Using tax to shift demand towards recycled plastic inputs.
- Using tax to encourage items to be designed in a way that makes them easier to recycle.
- Taxes or charges on specific plastic items that are commonly used on-the-go and littered, in order to encourage a reduction in production and use.
- Using tax to ensure that the right incentives are in place to encourage greater recycling of waste that is currently incinerated.

The environmental community was less enthusiastic about the 2018 Environment Plan and has described it as lamentably slow, and all commentators from the UK's NGO sector were in agreement that 25 years is not good enough (Chillcott, 2018). But an interesting idea is that the policy could be what picks up the 'laggards', as many retailers and companies have started to become innovative in response to the public pressure that has built around the issue. The UK Government is emphasising the business opportunities, saying that 'by making the most of emerging technologies, we can build a cleaner, greener country and reap the economic rewards of the clean growth revolution' (Defra, 2018:4). This rhetoric is repeated in the Chancellor of the Exchequer's introduction to the call for evidence to help the UK move to a greener economy 'ensuring the right incentives are in place to encourage sustainable behaviours and drive technological progress, which in turn will create new jobs and prosperity' (Treasury, 2018a:2). To ensure a successful economy, the 25-year Environment Plan and Treasury consultation encouraged businesses to adapt to less single-use plastics, whereby businesses can see an advantage while regulation will 'pick up the laggards' (Chillcott, 2018). The call from the Treasury to 'Tackle plastic pollution can therefore be seen as positive action by the government, to use the tax system or charges to address single-use plastic waste' (Treasury, 2018a).

Are big brands and the global plastics industry part of the solution or part of the problem?

Businesses have reacted differently to the government's 25-Year Environment Plan in different ways. Shortly after publication, the supermarket chain Iceland committed to eliminating plastic packaging (Slawson, 2018). The supermarket chain Morrisons advised that 'paper bags will replace plastic for loose fruit and vegetables' (Field, 2018), and there has been a lot of attention on the prospect of plastic-free aisles (Taylor,

2018). In response to these announcements, lobby groups such as the British Plastics Federation, the most powerful voice in the UK plastic industry with over 500 members across the plastics industry supply chain, including polymer producers and suppliers representing 80 per cent of the UK plastics industry, alerted retailers that these measures would not have any benefit for the marine environment and that their environmental footprints would instead increase (British Plastics Federation, 2018a), thereby discouraging the elimination of plastic packaging. The federation, together with Plastics Europe, is also engaged in ongoing lobbying against taxes on virgin polymers and single-use plastic products (British Plastics Federation, 2018b).

Since the public attention has turned, we have seen many large companies, including Coca Cola, Pepsi Co and Nestlé, who are responsible for generating the largest amounts of plastic packaging and waste found in the marine environment globally (Greenpeace, 2018), joining in the UK Plastics Pact with the aim of tackling the causes of plastic waste, not just the symptoms. In the media and the public, the plastics industry and big brands are now re-branding and presenting themselves as pro-active and part of the solution, which is a significant and laudable U-turn. Corporate social responsibility (CSR) initiatives that reduce plastic waste and sustainable plastics industry principles would be valuable initiatives that could have positive impacts (Landon-Lane, 2018). For example, Coca Cola is partnering with municipal governments to develop innovative solutions and initiatives, such as 'Print Your City', which converts plastic waste into useful items through 3D printing technology (Coca Cola, 2018).

However, prior to the *Blue Planet II* effect, the plastics industry was mostly involved in intense lobbying against any mandatory policies and targets to reduce plastic waste and bans on certain types of plastic, which has been well documented (Seydel, 2015). It is important to keep in mind not only the important role of key sectors such as plastics producers, but also that of retailers, the consumer goods industry, as well as importers, packaging firms and transport firms, and their influence on action to reduce plastic pollution. Another factor that is often forgotten, and which raises doubts about the sincerity of the corporate pledges and CSR initiatives, is the deep link between the plastics fossil fuel industries. All large oil companies like Shell, Chevron Phillips, Total and Sinopec own, operate, or are investing in plastics infrastructure (CIEL, 2017). Over 99 per cent of plastics are produced from chemicals sourced from fossil fuels and plastic consumption is important for the fossil fuel industry: if current trends were to continue, the consumption of oil by the entire plastics sector will account for 20 per cent of the total consumption by 2050 (CIEL, 2017). Fossil fuel subsidies also play a role as these incentivise the plastic market, allowing the cost of production to be less than the production of an alternative. As we know from the politics of climate change, the fossil fuel industries are a major obstacle to a low-carbon future. In the context of marine plastics, they appear to be obstructing both the reduction of petroleum-based plastics and eventually the complete replacement with renewable materials, particularly as the world begins to move away from fossil fuels. Are they now also becoming an obstacle to a zero waste future and the circular economy?

The circular plastics economy discourse

The discourse around the circular economy (CE), which increased substantially in the years prior to the marine plastics issue becoming a topic of public debate, also played a role in the recent UK policy initiatives. The Ellen MacArthur Foundation is referenced in the 25-Year Environment Plan. Ellen MacArthur herself advocated living within the finite resources of the planet in her TED talk in 2015 (MacArthur, 2015). The Ellen MacArthur Foundation has published numerous documents in recent years, starting with 'Towards the Circular Economy Vol 1–3: an economic and business rationale for an accelerated transition' (Ellen MacArthur Foundation, 2012; 2013; 2014) and more recently 'The New Plastics Economy: rethinking the future of plastics & catalysing action' (Ellen MacArthur Foundation, 2016), which the 25-year Environment Plan refers to.

However, according to Green Party leader Caroline Lucas, the circular economy has not been a major influence on what the government is doing with regard to plastics. Lucas is part of an all-party group that puts limits on growth, but she advised 'there aren't many allies, and it doesn't feel yet that we've managed to really push that out into other parts of government' (Chillcott, 2018). When referring to the CE, the 25-year plan argues that 'a healthy economy depends on a healthy environment' (Defra, 2018:84). Similar to the natural capital discourse being primarily to utilise natural resources for human consumption, the plan's use of the term CE revolves around efficiency to reduce waste and costs (ibid).

Further documents relating to the circular economy discourse with potential influence on the UK include 'Closing the Loop: an EU action plan for the circular economy' (European Commission, 2015). The circular economy was referenced in 61 debates between 2014 and summer 2018, not including sessions on 'topical questions'. The debates were counted and coded on topics, and issues were around waste and recycling; economics and industry; the EU and Brexit and the environment. The European Commission has been discussing the CE in line with its sustainability discourse (Ec.europa.eu, 2018) and the UK Parliament had a debate on the 'EU Action Plan for the Circular Economy' in March 2016 (Great Britain. House of Commons, 2016).

There is growing attention around a sustainable design for an economy, replacing the linear model of 'mass production and mass consumption' (Esposito, Tse and Soufani, 2018:6). Esposito, Tse and Soufani identify the arguments for a CE, which 'could potentially eliminate 100 million tonnes of waste globally in the next five years' (ibid). They describe the idea of shifting from a linear economy to a circular economy by stopping our single-use lifestyle. The idea of economic value, and the importance of this economic model to government policy, is also discussed, and they suggest that 'value creation continues to be critical in moving the circular economy from concept to practice' (ibid:13). The new model has the potential to be disruptive as well as innovative as it effects government policy, businesses and consumers (ibid:6). This is again both quantitative and qualitative evidence that UK Government policy on plastic waste is at least indirectly influenced by the current view of the CE.

The China factor and initiatives in the Global South

At the same time as the UK public and government were digesting the messages from *Blue Planet II* about plastic waste in oceans and food chains, China started the implementation of a new policy to ban the import of 24 kinds of plastic from the beginning of 2018. China is not only the largest plastics producer worldwide, accounting for 29 per cent of global production (Plastics Europe, 2018), it has also been a major player in the global plastics recycling value chains. The UK and other countries have relied on exporting plastic recycling to China for over 20 years. Many UK recycling businesses stopped shipping plastic to China already in the autumn of 2017 because of fears it might not arrive in time before the deadline. Through the ban, China has been causing disruption for many countries, including the UK, which have depended on China taking our 'Western garbage'. The impacts and ripples of this far-reaching policy have been felt all the way to the local level, where councils have been struggling to deal with accumulating amounts of plastic waste with no options and capacities for local recycling. Recent investigations have shown that the ban is costing local councils in England up to £500,000 extra a year, as they struggle to deal with the increasing amounts of plastic waste. (Laville, 2018).

As a result, plastic waste exports have shifted to other countries, in particular in South-East Asia, including Malaysia, Thailand and Vietnam. The countries saw significant increases in scrap plastic import volumes in 2018. UK plastic waste exports tripled over the first four months of 2018, according to HM Revenue and Customs (HMRC) data (Cole, 2018). Whilst the Chinese ban might help developing countries to build up their own recycling capacities, it runs the risk of overburdening these countries with negative impacts for environment, communities and oceans. As a result of the growing imports of plastic waste, several South-East Asian nations are now enacting plastic scrap import restrictions of their own, for example Thailand, following the Chinese example (Staub, 2018).

A major issue contributing to marine plastic pollution is the lack of waste management infrastructure in the developing world. The BBC documentary *Drowning in Plastic* from 2018 demonstrated how people living below the poverty line are purchasing large quantities of products wrapped in single-use plastic sachets. These products contain daily life necessities, such as washing power, on a smaller scale to make them affordable, but are individually wrapped in a material that cannot be disposed of correctly. In addition to the waste that developing countries create themselves, many import waste from the West, and as reported by the Guardian this year much of this ends up in uncontrolled landfills (Parveen, 2018). Both these issues are largely attributed to the limited or no infrastructure available to deal with plastic waste. A harrowing scene from *Drowning in Plastic* showed a community-made open landfill in a village, from which plastic waste overflowed into the waterways. The local people had used the same site for centuries, and traditionally discarded natural materials have been replaced by plastic in recent years.

Many governments in the Global South have already taken strong policy initiatives, like bans or levies on single-use plastics. The UN's recent 'Single-Use Plastics: a roadmap for sustainability' report highlights the fact that African countries are leading the way in the introduction and implementation of policies on plastic bags (UNEP, 2018). Some countries like Kenya have taken radical measures: Kenyans producing, selling or even using plastic bags will risk imprisonment of up to four years or fines of US$40,000 (£31,000). The controversial policy reflects the urgency of the crisis which is impacting the country. An example from western India is the state of Maharashtra, where one of the world's strictest plastic bans came into effect in June 2018. Penalties for manufacturing and selling single-use plastic items include fines of up to US$350 and jail terms of up to three months. However, these top-down measures have resulted in backlashes from industry and consumers alike (Chandrashekar, 2018). These examples show that it is not easy to simply restrict or ban a material that has become so deeply embedded in the modern economy, and top-down measures are not necessarily the best approach to enable a smooth and effective transition to more sustainable alternatives.

In contrast, research on community-based recycling and waste management in Pakistan's informal settlements demonstrates that bottom-up community-led initiatives are very effective: air pollution from the open burning of waste is estimated to cause 14,000 premature deaths a year in Pakistan, while openly dumped waste is a major cause of diarrhoeal diseases. A community-based approach to waste management in Islamabad addresses these problems while also creating jobs. A centre in piloting this approach provides US$10 in social, economic and environmental benefits for every dollar that was invested in establishing it. This approach reduces the need for more expensive, centralised waste management facilities by up to 90 per cent (Gower and Schröder, 2018).

The Global South is rich in innovative sustainability solutions, and this includes solutions to address plastics pollution (Nagendra et al., 2018). Solutions to alternative and sustainable packaging could possibly come from the oceans and the Global South. Seaweed is emerging as a promising new resource for biodegradable bioplastics. It is an abundant and versatile material and there is no major conflict with other uses; for example, there are no food security conflicts compared with other alternatives made from food crops such as corn, nor are there damaging effects on ecosystems. Indonesia, as the world's largest producer of red seaweed, whose carbohydrate element is the key ingredient in bioplastics, is in a prime position to lead this new field of packaging innovation (Sedayu, 2018). The Jakarta-based company Evoware is developing alternative packaging solutions from seaweed and this new emerging industry could provide many new economic opportunities for South-East

Asian countries. However, the issue of limited funding for science and technology R&D in developing countries could impact the innovation and development of the new materials and technologies (Krishna, 2001).

Conclusion: the role of transformative alliances for an international agreement on marine plastics pollution

Marine plastics pollution is a complicated, global, multi-dimensional problem that will require enormous political will to solve. Overcoming the problems and barriers to solutions, be they technical or social, is an inherently political process. It requires building coalitions and transformative alliances of reform-minded actors with the strength to change policy, create regulations or legislation, and – most importantly – implement them. In particular bans on unnecessary single-use plastics will require outmanoeuvring and overcoming powerful blockers such as the chemical and fossil fuels industries, if they cannot be persuaded to change their views and adjust their business interests and practices in view of the scientific facts. A positive example of the impact of transformative alliances is evidenced in the European Parliament's vote to cut pollution from single-use plastic items, which took place while we were finalising this chapter. The parliament voted to ban some of the most problematic throwaway products, such as expanded polystyrene food containers.

It is clear that a solution to marine plastics pollution requires the active participation of industry and business. But instead of relying on volatile global plastic recycling markets and vague voluntary commitments by the industry, becoming serious about extended producer responsibility (EPR) schemes is an important part of the solution. The new EU regulations aim to introduce mandatory EPR schemes for all packaging by 2025, as the effectiveness of voluntary schemes has been questionable. To ensure that producers are held accountable for the costs of single-use plastic pollution, the UK and other national governments need to show political will and follow these policy trends.

Regarding China's ban on plastic waste imports, which has caused some significant short-term disruption to waste management and recycling in the UK and other Western countries, in the long term, we would argue that China's action is having an ongoing positive influence on the UK's domestic policy, all the way down to the local municipal level. China's ban on foreign waste was clearly intended to develop and clean up the Chinese domestic waste infrastructure. Its far-reaching impacts raise the political question of whether the current free market approach and growing global trade in waste as a secondary resource is the right approach to tackle this global crisis. Following the initial ban, China has announced that other waste imports, including scrap steel, post-industrial plastics waste, polyethylene terephthalate (PET) bottles and e-waste, will be banned from entering China by the end of 2018. This should be a wake-up call for the UK and Europe that we seriously need to find local and national solutions to the crisis, rather than continue to export our waste to countries with inadequate waste management and recycling infrastructures.

To date, it has been the transformative alliances of civil society, activists, artists, film makers, ecologists, academics and socially responsible entrepreneurs which have been driving the current change in policy and pushing governments and companies to take action. At the local level, municipal governments need to become much more active in reducing plastic waste and in providing not only better local waste management services, but also integrative urban governance initiatives to reduce overall plastics use. Integrative approaches to plastic pollution will be able to achieve multiple targets of the SDGs, including Sustainable Cities and Communities (SDG 11), Sustainable Consumption and Production (SDG 12) and Life Below Water (SDG 13).

In order to move beyond short-term and localised responses, there is a very strong argument for an international agreement on marine plastic pollution by the UN. At the 2017 UN Environment Assembly, a global resolution on legally binding targets was opposed by major powers, including the US, China and India, resembling the dynamics of the international climate change negotiations. Only with continued action and growing alliances will it be possible to drive the change further, on local and global levels, tackling the complex politics of plastics pollution.

References

British Plastics Federation (2018a) British Plastics Federation's statement in response to the supermarket Iceland's announcement on plastic use, 18 January 2018. [Online] Available at: www.politicshome.com/news/uk/environment/environmental-protection/press-relea se/british-plastics-federation/92117/british

British Plastics Federation (2018b) Ministers reassure industry at British Plastics Federation parliamentary reception, 17 September 2018. [Online] Available at: www.bpf.co.uk/arti cle/ministers-reassure-industry-at-bpf-parliamentary-reception-1353.aspx.

Carson, H., Colbert, S., Kaylor, M. and McDermid, K. (2011) Small plastic debris changes water movement and heat transfer through beach sediments. *Marine Pollution Bulletin*, 62, 1708–1713.[Online] Available at: http://doi.org/10.1016/j.marpolbul. 2011.05.032

Chandrashekar, V. (2018) In India's largest city, a ban on plastics faces big obstacles. *Yale Environment 360*. Yale School of Forestry and Environmental Studies. [Online] Available at: https://e360.ya le.edu/features/as-indias-largest-city-shows-banning-plastics-is-easier-said-than-done

Chillcott, V. (2018) What influenced the UK Government to tackle plastic pollution? An analysis of the influences on public policy to tackle the problem of plastic pollution, particularly marine pollution, looking at the 25-year environment plan. Master's dissertation. Brighton: University of Sussex.

CIEL (2017) *Fuelling Plastics: fossils, plastics and petrochemical feedstocks*. Washington, DC: Centre for International Environmental Law.

Coca Cola (2018) How Coca-Cola can help make 'zero waste cities' a reality. *Coca Cola Company*, 2 October 2018. [Online] Available at: www.coca-colacompany.com/stories/ how-we-help-make-zero-waste-cities-a-reality

Cole, R. (2018) UK plastic exports to Malaysia triple following China ban. *Resource*, 19 June 2018. [Online] Available at: https://resource.co/article/uk-plastic-exports-malaysia-trip le-following-china-ban-12694

Dauvergne, P. (2018) Why is the global governance of plastic failing the oceans? *Global Environmental Change*, 51, 22–31.

Defra (2018) *A Green Future: our 25 year plan to improve the environment*. London: HM Government.

Dell, J. (2018) U.S. plastic recycling rate projected to drop to 4.4% in 2018. *Waste 360*. [Online] Available at: www.waste360.com/plastics/us-plastic-recycling-rate-projected-drop-44-2018.

Elgot, J. (2018) UK public backs tough action on plastic waste in record numbers. *The Guardian*. [Online] Available at: www.theguardian.com/environment/2018/aug/18/uk-public-backs-tough-action-on-plastic-waste-record-numbers-consultation-latte-levy-tax [Accessed 30 August 2018].

Ellen Macarthur Foundation (2012) *Towards the Circular Economy*. Volume 1. [Online] Available at: www.ellenmacarthurfoundation.org/assets/downloads/publications/Ellen-MacArthur-Foundation-Towards-the-Circular-Economy-vol.1.pdf [Accessed 30 August 2018].

Ellen Macarthur Foundation (2013) *Towards the Circular Economy*. Volume 2. [Online] Available at: www.ellenmacarthurfoundation.org/assets/downloads/publications/Ellen-MacArthur-Foundation-Towards-the-Circular-Economy-vol.1.pdf [Accessed 30 August 2018].

Ellen Macarthur Foundation (2014) *Towards the Circular Economy*. Volume 3. [Online] Available at: www.ellenmacarthurfoundation.org/assets/downloads/publications/Ellen-MacArthur-Foundation-Towards-the-Circular-Economy-vol.1.pdf [Accessed 30 August 2018].

Ellen MacArthur Foundation (2016) *The New Plastics Economy: rethinking the future of plastics & catalysing action*. [Online] Available at: www.ellenmacarthurfoundation.org/assets/downloads/publications/NPEC-Hybrid_English_22-11-17_Digital.pdf [Accessed 11 June 2018].

Ellen MacArthur Foundation (2017) *The New Plastics Economy: rethinking the future of plastics and catalysing action*. Isle of Wight, UK: Ellen MacArthur Foundation.

Esposito, M., Tse, T. and Soufani, K. (2018) Introducing a circular economy: new thinking with new managerial and policy implications. *California Management Review*, 60 (3), 5–19.

European Commission (2014) *Final Report: development of guidance on extended producer responsibility (EPR)*. Paris: Deloitte.

European Commission (2015) *Closing the Loop: an EU action plan for the circular economy. COM/2015/0614 final*. Brussels: European Commission. [Online] Available at: https://eur-lex.europa.eu/legal-content/EN/TXT/?uri=CELEX:52015DC0614.

European Commission (2018) *A European Strategy for Plastics in a Circular Economy. COM (2018) 28 final*. [Online] Available at: http://ec.europa.eu/environment/circular-economy/pdf/plastics-strategy.pdf

Evening Standard (2018) Brexit. [Online] Available at: https://www.standard.co.uk/topic/brexit [Accessed 27 August 2018].

Field, M. (2018) Morrisons brings back paper bags for groceries to cut out plastic. *The Telegraph*. [Online] Available at: https://www.telegraph.co.uk/business/2018/06/24/morrisons-brings-back-paper-bags-groceries-cut-plastic/ [Accessed 27 August 2018].

GESAMP (2016) Sources, fate and effects of microplastics in the marine environment: part 2 of a global assessment. In Kershaw, P.J. and Rochman, C.M. (Eds.) *IMO/FAO/UNESCO-IOC/UNIDO/WMO/IAEA/UN/UNEP/UNDP Joint Group of Experts on the Scientific Aspects of Marine Environmental Protection*. Report Studies GESAMP No. 93.

Geyer, R., Jambeck, J. and Law, K. (2017) Production, use, and fate of all plastics ever made. *Science Advances*, 3 (7), e1700782. [Online] Available at: https://doi.org/10.1126/sciadv.1700782.

Gower, R. and Schröder, P. (2018) *Cost-Benefit Assessment of Community-Based Recycling and Waste Management in Pakistan*. London and Brighton: Tearfund and Institute of Development Studies. [Online] Available at: https://learn.tearfund.org/~/media/files/tilz/circular_economy/2018-tearfund-cost-benefit-assessment-pakistan-en.pdf?la=en.

GOV.UK (2018) Natural Capital Committee (NCC). [Online] Available at: www.gov.uk/government/groups/natural-capital-committee [Accessed 27 August 2018].

Great Britain. House of Commons (2016) *The Official Report: parliamentary debates* (Hansard), 7 March. London: The Stationery Office. [Online] Available at: https://hansard.parliament.uk/Commons/2016-03-07/debates/a6a765b8-0aa9-4258-a19b-5187c445b4eb/EUActionPlanForTheCircularEconomy?highlight=%22circular%20economy%22#contribution-16030711000001[Accessed 20 July 2018].

Great Britain (2018) Hansard. [Online] Available at: https://hansard.parliament.uk/ [Accessed 30 August 2018].

Greenpeace (2018) Coca-Cola, PepsiCo, and Nestlé found to be worst plastic polluters worldwide in global cleanups and brand audits. Press release, 9 October 2018. Amsterdam: Greenpeace International.

The Guardian (2018) Brexit. *The Guardian*. [Online] Available at: https://www.theguardian.com/politics/eu-referendum [Accessed 27 August 2018].

Jambeck, J., Geyer, R., Wilcox, C., Siegler, T., Perryman, M., Andrady, A., Narayan, R. and Lavender Law, K. (2015) Plastic waste inputs from land into the ocean. *Science*, 347, 768–771. [Online] Available at https://doi.org/10.1126/science.1260352.

Krishna, V. (2001) Changing policy cultures, phases and trends in science and technology in India. *Science and Public Policy*, 28 (3), 179–194.

Landon-Lane, M. (2018) Corporate social responsibility in marine plastic debris governance. *Marine Pollution Bulletin*, 127, 310–319.

Laville, S. (2018) Plastic recycling industry's problems costing councils up to £500,000 a year. *The Guardian*, 20 October 2018. [Online] Available at: www.theguardian.com/environment/2018/oct/20/plastic-recyclings-problems-costing-councils-up-to-500000-a-year.

Leous, J. and Parry, N. (2005) Who is responsible for marine debris? The international politics of cleaning our oceans. *Journal of International Affairs*, 59 (1), 257–269.

Mail Online (2018) Viewers criticise Blue Planet II for linking plastic to whale death. [Online] Available at: www.dailymail.co.uk/news/article-5100643/Blue-Planet-II-slammed-linking-plastic-dead-whale.html [Accessed 30 August 2018].

MacArthur, E. (2015) The surprising thing I learned sailing solo around the world. *Ted.com*. [Online] Available at: www.ted.com/talks/dame_ellen_macarthur_the_surprising_thing_i_learned_sailing_solo_around_the_world [Accessed 27 August 2018]

Mendenhall, E. (2018) Oceans of plastic: a research agenda to propel policy development. *Marine Policy*, 96, 291–298. [Online] Available at: https://doi.org/10.1016/j.marpol.2018.05.005

Nagendra, H., Bai, X., Brondizio, E. and Lwasa, S. (2018) The urban south and the predicament of global sustainability. *Nature Sustainability*, 1, 341–349.

Parveen, N. (2018) UK's plastic waste may be dumped overseas instead of recycled. *The Guardian*. [Online] Available at: www.theguardian.com/environment/2018/jul/23/uks-plastic-waste-may-be-dumped-overseas-instead-of-recycled [Accessed 17 October 2018].

Plastics Europe (2018) Plastics – the facts 2017. Brussels: Plastics Europe. [Online] Available at: www.plasticseurope.org/application/files/5715/1717/4180/Plastics_the_facts_2017_FINAL_for_website_one_page.pdf

Rawlinson, K. (2017) Michael Gove 'haunted' by plastic pollution seen in Blue Planet II. *The Guardian*. [Online] Available at: https://www.theguardian.com/environment/2017/dec/19/michael-gove-haunted-by-plastic-pollution-seen-in-blue-planet-ii [Accessed 27 August 2018]

Schmidt, C., Krauth, T. and Wagner, S. (2017) Export of plastic debris by rivers into the sea. *Environmental Science & Technology*, 52, 12246–12253.

Sedayu, B. (2018) Seaweed, Indonesia's answer to the global plastic crisis. *The Conversation*, 4 June 2018. [Online] Available at: https://theconversation.com/seaweed-indonesias-answer-to-the-global-plastic-crisis-95587 [Accessed 23 December 2018].

Seydel, L. (2015) Powerful lobbying groups want to make sure you keep using plastic bags. *Huffington Post*, 15 October 2015. [Online] Available at: www.huffingtonpost.com/laura-turner-seydel/powerful-lobbying-groups-want-to-make-sure-you-keep-using-plastic-bags_b_8307416.html [Accessed 23 December 2018]

Silver, J., Gray, N., Campbell, L., Fairbanks, L.and Gruby, R. (2015) Blue economy and competing discourses in international oceans governance. *The Journal of Environment & Development*, 24 (2), 135–160.

Slawson, N. (2018) Iceland supermarket vows to eliminate plastic on all own-branded products. *The Guardian*. [Online] Available at: www.theguardian.com/business/2018/jan/15/iceland-vows-to-eliminate-plastic-on-all-own-branded-products [Accessed 27 August 2018].

Staub, C. (2018) Basel amendment could further slow U.S. plastic exports. *Resource Recycling*, 1 August 2018. [Online] Available at: https://resource-recycling.com/plastics/2018/08/01/basel-amendment-could-further-slow-u-s-plastic-exports/

The Sun (2018) Brexit. [Online] Available at: https://www.thesun.co.uk/topic/brexit/ [Accessed 27 August 2018].

Taylor, M. (2018) World's first plastic-free aisle opens in Netherlands supermarket. *The Guardian*. [Online] Available at: www.theguardian.com/environment/2018/feb/28/worlds-first-plastic-free-aisle-opens-in-netherlands-supermarket [Accessed 27 August 2018].

Treasury (2018a) *Tackling the Plastic Problem: using the tax system or charges to address single-use plastic waste*. London: HM Government.

Treasury (2018b) *Tackling the Plastic Problem: summary of responses to the call for evidence*. London: HM Government.

UNEP (2014) *Valuing Plastic: the business case for measuring, managing and disclosing plastic use in the consumer goods industry*. Nairobi: United Nations Environment Programme.

UNEP (2018) *Single-Use Plastics: a roadmap for sustainability*. Nairobi: United Nations Environment Programme.

Villarrubia-Gómez, P.Cornell, S. and Fabres, J. (2017) Marine plastic pollution as a planetary boundary threat: the drifting piece in the sustainability puzzle. *Marine Policy*, 96, 213–220. [Online] Available at: https://doi.org/10.1016/j.marpol.2017.11.035

4

CIRCULAR ECONOMY AND INCLUSION OF INFORMAL WASTE PICKERS

Political economy perspectives from India and Brazil

Patricia Noble

Introduction

The issue of resource mismanagement resulting in an increasing quantity of solid waste can be attributed to our linear economy of 'take, make, dispose'. While historically, these problems have been caused by the now developed countries, the challenges have also emerged for developing economies which are following the linear economic development trajectory of countries in the Global North. The alternative development approach is a circular economy (CE), which, so far, remains less understood in the context of developing countries. In low and middle-income countries, many CE activities exist in the form of waste picking in the informal sector. About 1 per cent of the urban population in developing countries is involved in waste picking activities (Medina, 2008) – based on an urban population which reached over 3 billion in developing countries in 2016 (UNCTAD, 2017), this would be more than 30 million people working as informal waste pickers (IWP) worldwide. These individuals and groups are pivotal for 'closing the loop' of product lifecycles in the CE. This is accomplished through the process of collection, transportation and transformation of waste into valuable secondary resources. As a result, municipal waste is reduced and resources are inserted back into the economy – also referred to as closing the waste management cycle (WMC).

In order to understand the enabling conditions that allow IWP to recycle and effectively contribute to closing the WMC, this chapter explores the conditions under which IWP operate in developing countries. This is approached through an in-depth literature review based on the case studies of Delhi, India, and the São Paulo Metropolitan Area (SPMA), Brazil, by identifying the various stakeholders that produce different narratives and pathways, contributing to the Green Transformations of cities.

Connecting the circular economy to Green Transformations

The concepts and approaches of recycling, the WMC, CE and Green Transformations are interconnected as the recycling of waste enables the creation of a closed looped municipal waste system. These activities allow for the transition from a linear model to a circular model, forming part of the wider concept of Green Transformations (Scoones, Leach and Newell, 2015), as depicted in Figure 4.1. The transition from a linear economy to a CE forms part of the wider discussion of Green Transformations, as it aims to benefit society by meeting current demand whilst simultaneously reducing the environmental impact. A fundamental aspect of the CE is recovering resources and retaining the value that can be found in products by bringing them back into the production and consumption processes of the economy through practices of repair, reuse and recycle (Chaturvedi, Vijayalakshmi and Nijhawan 2015).

A key arena where the CE has come to light is through closing the loops of the WMC. Most recently, the European Commission has put the CE at the heart of their waste management solution, where they have revised legislative proposals on waste, as part of their 'Circular Economy Package' (European Commission, 2017). However, in developing countries, particularly Brazil and India, WMC activities are predominantly undertaken in the form of collection and recycling by IWP in large urban areas.

Understanding informal waste pickers

IWP are commonly defined as individuals, from marginalised social groups of the urban poor, who perform waste collection and recycling activities to obtain a small

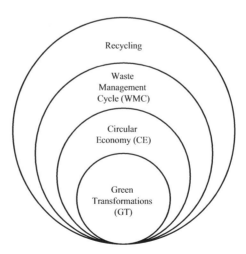

FIGURE 4.1 Nested systems: from recycling to Green Transformations.

source of income in order to ensure their survival (Moreno-Sanches and Maldonado 2006; Wilson, Velis and Cheeseman, 2006). Moreover, they are highly skilled in identifying materials of value in the municipal waste stream and locating customers within the market (Nzeadibe, 2009).

A significant body of literature has unveiled the contributions of IWP to closing the WMC, despite being considered a marginalised group and using simple techniques (Schindler, Demaria and Pandit, 2012; Chaturvedi, Vijayalakshmi and Nijhawan, 2015). IWP, especially in low and middle-income countries, recover large quantities of waste and divert significant quantities of material away from the waste stream, thus preventing materials ending up in landfill, open dumps or being openly burned. It was found that more than 80,000 people in the informal sector were involved in valourising around 3 million tonnes per year of waste across six cities in developing countries (Scheinberg, 2010b). Evidence indicates that these individuals have shown to be 'CE pioneers', especially in the 'labour intensive process of collection, manual segregation and dismantling of waste' (recycling) (Agarwal et al., 2002:X) as well as the reuse, redistribution and refurbishment of waste (Rogerson, 2001; Gerdes and Gunsilius, 2010; Chi et al., 2011; Nzeadibe, 2009; Gunsilius, Chaturvedi and Scheinberg, 2011; Gunsilius et al., 2011; Lines et al., 2016; Lines and Garside, 2014).

Moreover, the body of literature suggests that IWP are more effective in waste recovery than the formal sector, as shown in Table 4.1.[1] Having said this, the formal and informal sector are often linked in matters of production, consumption or distribution (Katusiimeh, Burger and Mol, 2013). A study conducted by the Informal Economy Monitoring Study in 2012[2] found that 76 per cent of formal businesses are the main buyers of segregated and recovered waste materials from IWP (Dias and Samson, 2016).

On the other hand, some disregard the contribution of IWP towards the CE, as it is believed that these individuals and groups will reduce efficiency when

TABLE 4.1 Comparison of waste recovery rates in seven cities.

	Belo Horizonte (Brazil)	Canete (Peru)	Delhi (India)	Dhaka (Bangladesh)	Managua (Nicaragua)	Moshi (Tanzania)	Quezon City (Philippines)
Tonnes per year recovered all sectors	145,135	1,412	841,070	210,240	78,840	11,169	287,972
% recovered by formal sector	0.1%	1%	7%	0%	3%	0%	8%
% recovered by informal sector	6.9%	11%	27%	18%	15%	18%	31%

Source: Gupta, 2012[3]

transitioning to a CE. This is said to be due to slowing down the value chain, as they are believed to contribute to leakages of material streams, resulting in inefficient reprocessing (World Economic Forum, 2014).

An investigation of the body of literature shows that there is a clear domination of reports, journals and articles regarding CE efforts and systems in more developed countries than those in the developing world.[4] This chapter aims to contribute to the gap in the literature by understanding how developing countries can transition, or even leapfrog, to a CE with the aid and inclusion of IWP.

Enabling conditions of IWP to recover and recycle

The following section identifies the specific processes by which IWP have been included in the waste management systems of the two cities of Delhi and Sao Paolo.

Case Study 1: Delhi, India

Background

India has transformed from 'a recycling to a throwaway society' (Schindler, Demaria, and Pandit, 2012:18).[5] This trend is led by major cities like Delhi: in 2015, across its five municipal authorities, the city generated approximately 9,620 tonnes of MSW per day (DPCC, 2015). In Delhi, waste management is controlled by formal public sector agencies, including the Municipal Corporation of Delhi, the New Delhi Municipal Council (NDMC), and the Delhi Cantonment Board (Chintan Environmental Research and Action Group, 2009).

The collection and recycling rates in waste management by the informal sector are responsible for about 27 per cent of waste recovery in Delhi (Gupta, 2012). However, many of the estimated 150,000–200,000 waste workers (Chaturvedi and Gidwani, 2011) belong to marginalised and vulnerable communities that struggle to find alternative livelihoods (Gill, 2010). The workforce in Delhi consists of IWP, small middlemen (small *kabaris*), collectors (*thiawalas*), and big middle-men (big *kabaris*), with IWP being the largest and most vulnerable of all (see Figure 4.2) (Chintan Environmental Research and Action Group, 2009). This being said, Delhi is known as one of the prime locations for e-waste disposal and recycling in India, where the informal sector manages most of the waste in these areas (Sthiannopkao and Wong, 2013; Pradhan and Kumar, 2014). IWP generally segregate waste in *dhalaos* (neighbourhood disposal units, resembling concrete sheds), in addition to collecting recyclables from the street, public bins, door-to-door collection and landfills (Chintan Environmental Research and Action Group, 2009).

Actors and roles

One of the main conditions that support IWP in collecting and recycling are waste picker organisations. From the 1990s onwards Delhi, amongst other cities, has

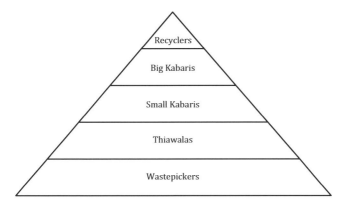

FIGURE 4.2 The informal recycling system in Delhi.
Source: Chintan Environmental Research and Action Group, 2009, modified by author.

unionised IWP into cooperatives, with the understanding that their livelihoods can be protected and enhanced by fostering the segregation of waste at its source and door-to-door collection to guarantee access to scrap, improving working conditions and earnings, as well as allowing them to obtain the status of service provision (Gerdes and Gunsilius, 2010). Consequently, in 2005 the National Alliance of Waste Pickers was founded, allowing for the inclusiveness of IWP in solid waste management (Gerdes and Gunsilius, 2010).

Since 1972, NGOs have facilitated conditions for the integration of IWP in India. In Delhi, the NGOs SriFsti and Chintan Environment Research and Action Group have played an important role in enabling private partnerships, training waste collectors, and supporting the government of Delhi to spread awareness of the benefits of IWP through awareness generation programmes, school programmes and door-to-door meetings (Taylan, Dahiya and Sreekrishnan, 2008). More specifically, the NGO Chintan Environment Research and Action Group has organised around 14,500 IWP with the aim of being in partnership with the recyclers as well as Municipal Corporation of Delhi and the New Delhi Municipal Council to promote door-to-door collection (Samson, 2009). Through this, IWP were able to mobilise themselves to demand contracts for door-to-door collection and after several weeks a formal agreement was signed allowing the right for WPs to collect from approximately 50,000 households. Chintan identified the active engagement of the workers as a vital component in the success of this initiative, ultimately empowering marginalised groups. As a result, the New Delhi Municipal Council enabled conditions for IWP by: (1) easing of work through sanitary inspectors, (2) providing space for waste segregation, (3) assisting access medical facilities and distributed identity cards for WPs though Chintan, and (4) identifying institutional obstructions that involved IWP. In only a few months there was a reduction in verbal abuse, bribes and forced labour by municipal officials. Surveys further highlighted the success of door-to-door collection as areas were not only cleaner, but also waste was disposed every day, allowing for regular recycling (SNDT Women's University & Chintan Environmental Research and Action Group, 2008).

Finance and infrastructure

Informal recycling cooperatives in Delhi have also responded to market demands for secondary raw materials, which have emerged over the years (Köberlein, 2003). Studies suggest that IWP have obtained higher incomes through market-led capabilities (Samson, 2009). For example, in the partnership described previously with NGO Chintan Environment Research and Action Group and the door-to-door collection initiative, IWP were initially prevented by the New Delhi Municipal Council (NDMC) from charging a fee to residences for their services. As a result, IWP requested a small financial contribution from households towards a cup of tea every week, in order to obtain a modest income (Samson, 2009). Through the removal of obstacles presented by the municipality, IWP were allowed to charge for their services, so that 70 per cent of the clients now pay the IWP for collection (Samson, 2009). Moreover, a 2008 study revealed that the income of IWP had increased significantly from $59–$71 per month to $126 per month due to the increased number of residents signing up for door-to-door collection (SNDT Women's University & Chintan Environment Research and Action Group, 2008).

It is important to identify not only the enabling conditions, but also those conditions that inhibit IWP from recycling. In 2005, the government of Delhi took steps shifting towards the privatisation of waste management involving the formal sector, including neighbourhood collection points and transport to landfills. This barred IWP from accessing waste via landfills and bins, where an informal social sharing was visible (SNDT Women's University & Chintan Environment Research and Action Group, 2008; Gidwani and Reddy, 2011). Moreover, the privatisation of *dhalaos* hindered recycling, as IWP were unentitled to their 'work space' to effectively segregate waste after engaging in door-to-door collection activities (SNDT Women's University & Chintan Environment Research and Action Group, 2008; Samson, 2009). As a result, these individuals had to travel greater distances in order to dispose of household waste, and were subjected to physical punishment when they did not comply with the new privatisation rules (SNDT Women's University & Chintan Environment Research and Action Group, 2008).

Governance and capacities

There is a sense of growing acknowledgment of the informal recycling cooperatives in national policies, including the National Environment Policy in 2006[6] as well as the National Action Plan for Climate Change in 2009, which also refers to IWP (SNDT Women's University & Chintan Environment Research and Action

Group, 2008). Nonetheless, the municipality in Delhi has failed to take this into consideration when privatising waste management and establishing public-private partnerships (PPPs) with the formal sector, as already discussed (Chaturvedi, Arora and Saluja, 2015). In the Municipal Corporation of Delhi area, approximately 50 per cent of IWP have reported lower incomes and job losses, resulting in relocation to other nearby areas and thus impacting the income of IWP in those area that had not yet become privatised (SNDT Women's University & Chintan Environment Research and Action Group, 2008).

Case Study 2: São Paulo Metropolitan Area (SPMA), Brazil

Background

In recent decades the increase of the urban population as well as changes in consumption rates have made it difficult for pre-existing waste disposal infrastructure to cope with the vast volumes of solid waste produced (Consonni, 2013). However, Brazil has exemplified itself in terms of waste picker inclusion and has become internationally renowned for its recycling levels (Rutowski and Rutowski, 2015).

Brazil's municipalities are officially responsible for selective waste collection services. Nonetheless, only 18 per cent of municipalities have formal selective waste collection programmes as most materials are recovered by informal collectors and recyclers (IPEA, 2012). The municipalities often outsource or are in partnership with waste picker associations and cooperatives that recover waste (IPEA, 2010; BRASIL, 2013). In a global comparison, Brazil has the highest number of ventures with IWP, at 1,000 amongst the 5,500 municipalities nationwide (BRASIL, 2011).

Brazil's waste management is characterised by its integration of IWP, named *catadores*, and the informal recycling system. In Brazil, 387,910 individuals stated they were IWP, and in the state of São Paulo alone, 79,770 individuals identify as IWP, equivalent to 20 per cent of the national total (Besen et al., 2014), which is where the largest concentration of waste pickers in the country is found (Gunsilius et al., 2011).

Actors and roles

Social activism has a strong presence in Brazil's history. The focus on IWP in waste management can be attributed to the efforts made by the Catholic Church and its associated NGOs through socio-pedagogical work. In the 1970s, the Catholic Church initiated projects with a focus on street dwellers, developing the first waste picker cooperative in 1980 (Dias and Alves, 2008; Gerdes and Gunsilius, 2010; Van Zeeland, 2014). In São Paulo, the first waste picker cooperative was established in 1989 (Gerdes and Gunsilius, 2010). Since then, IWP in Brazil have orchestrated themselves, with the support of NGOs, to form associations or cooperatives, leading to the inclusion of their work in municipal waste management structures and improving their livelihoods, which has now been referred to as a global best practice (Dias, 2011; Van Zeeland, 2014).

The collective action of waste picker cooperatives and NGOs nationwide exhibits the social power and influence on public policies. In 2001, Brazil's national waste picker movement Movimento Nacional dos Catadores de Materiais Recicláveis (MNCR) was established.[7] The involvement of over 1,600 IWP in the national congress prompted the *Carta Brasília* (Letter of Brasilia), reinforcing the recognition and regulation of the waste picker vocation, and in 2002 the federal government officially recognised the profession of *Catador de Material Reciclável* (collector of recyclable material) (Van Zeeland, 2014).

Cooperatives have become involved in multiple stakeholders platforms, as well as become formally organised, empowering their rights as economic actors in solid waste management. Moreover, cooperation and partnership between formal authorities, the private sector and waste picker cooperatives have enabled the facilitation and inclusion of waste pickers in the recycling sector (Gerdes and Gunsilius, 2010). Over the years, waste picker cooperatives have developed a social technology (Rutkowski and Lianza, 2004; Dagnino, 2006), called Solidarity Selective Collection (*Coleta Seletiva Solidariedade* [CSS] in Portuguese) (Rutkowski and Rutkowski, 2015). Through this, waste picker cooperatives are contracted by municipalities as formal operators in the solid waste management system. The CSS initiative educates communities on recycling practices and requests that households segregate and store the recyclable materials separately from waste. The citizens are informed on the importance of recycling and segregation and the impact these make on the livelihoods of waste pickers (Rutkowski and Rutkowski, 2015). This approach has focused on the broader social and economic benefits of recycling towards consumers, as Brazil is more inclined to recycle for the livelihoods of the poor than for environmentally conscious reasons (Rutkowski and Rutkowski, 2015; Wheeler and Glucksmann, 2015).

Moreover, through solidarity, IWPs are able to respond to market needs. The recyclables collected are segregated into types of material in the storage sheds, which are then sold to intermediaries and then on to the industry. The organisation of solidarity commercialisation networks, where several waste pickers work together to sell volumes of materials as a single entity, enable them to respond more effectively to market conditions and industry requirements e.g. high volumes of materials (Rutkowski and Rutkowski, 2015). These networks promote dialogue between waste pickers and the recycling industry as well as partnerships between formal private companies and public sector organisations, as waste picker cooperatives have access to financial resources as well as technical and managerial skills (Rutkowski and Rutkowski, 2015).

Finance and infrastructure

The involvement and collaboration of the state regarding waste picker cooperatives enables individual waste pickers to obtain a better standard of living and more incentives to recycle. Laws, such as that of President Lula de Silva to finance cooperatives and provide education and low-income housing, have shown the engagement of the state in including waste pickers in municipal waste management (De Brito, 2012). In 2004, Diadema (located in the São Paulo Metropolitan Area) became the first municipality to adopt a specific law, namely #2336/04 *Governing the System for the Sustainable Management of Solid Waste* (Dias and Alves, 2008), which authorises contracts with cooperatives to pay for their services as part of the municipal recycling programme (Samson, 2009). In 2005, the cooperatives were being paid the equivalent per tonne of recyclables to that of a private company (Dias and Alves, 2008). Through this legislation, the city has created 60 locations to which residents can take their recyclables. The income of the *catadores* has increased due to the additional payment of the municipality contracts as well as the selling of recyclables. Additionally, as groups, they can obtain higher prices for their materials by cutting out the middleman and selling directly to industries (Gutberlet, 2008).

Regarding the infrastructure, it was noted that the cooperatives were 'well established' via internal organisation, working spaces, trucks and machinery as well as support from NGOs, municipalities and development banks (Marello and Helwege, 2014:14). Public areas and commercial spaces have been set up by municipalities to provide an area to segregate recyclable materials (Riberio et al., 2009). However, most waste pickers still face difficulty in fulfilling their recycling capabilities due to a lack of space. If more space were provided, cooperatives could hire more people, increasing employment as well as recycling waste (Marello and Helwege, 2014).

Governance and capabilities

In the case of Brazil, the state has shown responsiveness to the pressure for social inclusion as a result of the social movements of waste pickers. Additionally, the Brazilian government has shown activeness in progressive legislation for the inclusion of informal recyclers through public policies and laws, especially for cooperatives regarding recyclable collectors (Dias, 2009).[8]

Arguably, one of the pivotal movements to strengthen waste pickers' organisation through recognition by the public and higher officials was the social-government multi-stakeholder platform known as the National Waste and Citizenship Forum in 1998. The movement, led by UNICEF, advocated a multi-stakeholder approach to encourage a network of state and municipal waste and citizenship forums throughout the country, in addition to the creation of national training programmes and the coordination of financing institutions to take into account the inclusion of pickers and waste disposal. The creation of the national forum gave credibility to the conditions of IWP (more specifically, children waste pickers) in the waste sector and their role in an integrated approach to urban waste. Their national campaign, named 'no more children in dump areas', reached unprecedented heights due to the attention and coverage of the media as well as the involvement of UNICEF's ambassador to Brazil. The campaign also involved sending a questionnaire and letter of commitment to 5,507 mayors across the nation for them to adhere to the objectives of the campaign. This movement, which was later passed onto the NGO Água e Vida, highlights the responsiveness of the state, as 25 per cent of mayors returned their letter of commitment. Amongst the numerous accomplishments of the forum, the State Secretariat for the Environment set out financial incentives for municipalities that support associations of waste pickers. The inclusion of IWP was also embedded into the mandate, whereby municipalities that close an open dump must create alternative projects for the IWP (Dias, 2006).

Brazil's first clear policy regarding recycling and solid waste management was the National Policy for Solid Waste (in Portuguese: *Politica Nacional de Residuos Solidos*), which came about in 2010 (IPEA, 2012). This policy advocates involving organised IWP in the process of selective waste collection. Although urban solid waste management (SWM) is overlooked by the municipal government, this national policy has created mechanisms to further facilitate selective waste collection by providing resources to municipalities who proceed with their integrated waste management plans in accordance with this policy (Besen et al., 2014). The selective waste collection initiative has been shown to enhance the inclusion of IWP, as a 2008 study by the National Basic Sanitation Survey revealed that a majority (65.7 per cent) of municipalities were offering selective waste collection with the involvement of organised IWP (IBGE, 2010). Additionally, this policy adopted the concept of 'reverse logistics', whereby all products from the production chain which have entered the economy from external sources are returned to the economy through recycling and reprocessing by the importers, producers and manufacturers

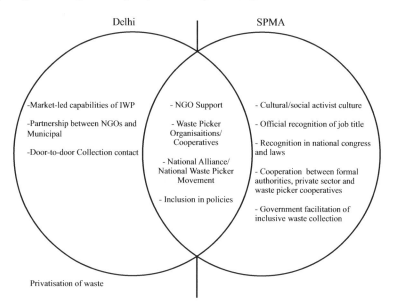

FIGURE 4.3 Venn diagram highlighting the main differences and similarities of the conditions that enable IWP to close the WMC via recycling in Delhi and the SPMA.

responsible. These materials can be reused as raw materials, thus removing the strain on raw resources and forming a CE (PNRS, 2012; Wheeler and Glucksmann, 2015). This has been undertaken in São Paulo, where municipals are obliged to hire waste picker collectives for reverse logistics in order to be eligible for federal resources for solid waste management (PNRS, 2012). This process enables dialogue and partnership between the government, the private sector and IWP.

Through the analysis it becomes clear that certain factors catalyse IWP's contribution to recycling, and thus to closing the WMC. The main similarities and differences are summarised in the Venn diagram presented in Figure 4.3.

Discussion and conclusion

Actors and roles

Both cases analysed showed a clear bottom-up approach that drives IWP to recover and recycle materials, thus closing the WMC, enabling a transition from the linear economic system of municipal waste to a more circular urban system. The acceleration can be seen as emerging from a local level in the Global South, via civic actors including NGOs as well as the IWP themselves, and the Catholic Church in the case of Brazil. One can assume that this was motivated by green jobs, or more likely, jobs themselves in order for IWP to obtain and support their livelihoods.

A citizen-led narrative is manifest in both cases where the cumulative actions of initiatives and networks can be seen through the aid of NGOs and other civic actors in order for IWPs to organise and orchestrate into waste picker cooperatives (Samson, 2009; Gerdes and Gunsilius, 2010; Van Zeeland, 2014; Dias, 2011). This approach has further allowed the creation of national waste picker alliances and movements, as well as their inclusion in policy and laws, emphasising the power and political economy of these grassroot initiatives (Roth, 2014; Gerdes and Gunsilius, 2010).

However, within the two case studies it can be seen that different incentives shape different narratives. The SPMA case study portrayed a more grassroots and citizen-led approach than the Delhi case, due to the deep-rooted origin of social activism seen through the support of the Catholic Church, NGOs and civil society. Additionally, the case of SPMA also portrays a relationship between citizen-led and state-led narratives through social-governmental processes, where on the one hand waste picker cooperatives were being created in the 1980s and, on the other hand, the responsiveness from the state is visible through the passing of supportive laws and legislation. Although there is a clear marketisation pathway due to the nature of collecting and selling secondary materials, the inclusion of IWP in SWM is still more socially led. This is emphasised through the citizen-led pathway carved by households and society to happily aid IWP through the household segregation of waste and recyclables. Additionally, to attract the involvement of the society in regards to the CSS, the government used a campaign to highlight the impact of recycling on the lives of IWP as Brazilians are more inclined to collaborate if they know they are making a difference towards somebody's life (Rutkowski and Rutkowski, 2015; Wheeler and Glucksmann 2015). This was seen as a more effective strategy than enforcing penalties, exhibiting the socially driven culture. On the other hand, although Delhi experiences a citizen-led approach through its support from NGOs and has shown collaboration through these programmes with the municipality, a more marketisation narrative is visible. This is due to the more economically incentivised push to respond to market demands for secondary raw materials (Köberlein, 2003) and through door-to-door collection contracts and market-led capabilities. Having said this, it is important to note that when comparing SPMA and Delhi, one also needs to analyse the nature of the modernisation of solid waste management in both countries. For decades, Brazil's household solid waste system has been considered to be highly modernised, in comparison to India, where this practice is not widespread. In India's scenario, IWP fill the gap in the absence of municipal waste collection, whereas in Brazil IWP are able to work in the collection and selling of recyclables due to the level of public recognition achieved by the state and their securement of co-production arrangements, in which they provide services in selective waste collection (Dias, 2009). This highlights the marketisation pathway created by a gap in the municipal SWM systems.

Moreover, the two cases allow for an in-depth understanding of the actors necessary to facilitate IWP contributions towards closing material loops. Scoones, Leach and Newell (2015:177) state that the 'no single actor has the resources to bring about the Green Transformations'. This is evident in the case of SPMA where many actors including public authorities, the private sector and civil society are involved. Through this multi-stakeholder cooperation, IWPs are provided with enabling conditions that facilitate recycling of waste. This supports the argument that there is a need to combine bottom-up and top-down approaches (Bulkeley and Newell, 2010; Scoones, Leach and Newell, 2015), and the vertical categorisations of actors to accelerate Green Transformations. In the case of Delhi, apart from collaborations between NGOs and waste picker cooperatives, which then partner with the municipalities, it is felt that the actors do not fully collaborate as one entity. This will be further discussed in the following section.

Finance and infrastructure

The Green Transformation narratives can be clearly identified through the observation of conditions that enable finance and infrastructure to support IWP in recycling activities. As expressed in the previous, Delhi presents a clear marketisation narrative of IWP and recycling through their market-led capabilities in order to obtain financial means, for example via their strategy to obtain a contribution from households towards a cup of tea, when it was not allowed for IWP to be paid for their door-to-door collection services. However, the NDMC then facilitated the access to finance by allowing the IWP to charge a fee as well as obtain contracts for the door-to-door collection services, through which their income grew significantly (Samson, 2009). This illustrates a pathway facilitated by a state-led narrative.

Having said this, the state has also implemented obstacles to achieving a Green Transformation through the privatisation of waste by the municipality (SNDT Women's University & Chintan Environment Research and Action Group, 2008; Gidwani and Reddy, 2011). Privatisation hinders enabling frameworks and incentives for IWP to recycle as they are unable to use the infrastructural 'work space' and their access to waste is challenged (SNDT Women's University & Chintan Environment Research and Action Group, 2008; Samson, 2009). This emphasises the need to include all three actors (public, private and civic) in order to achieve a successful Green Transformation, which has not been the case in Delhi, resulting in a decrease of recycling rates by IWP. Moreover, it supports the notion that 'attention needs to focus on supportive alliances across all categories' as well as to 'recognise that within the government, civil society, businesses there are actors that block or slow down the Green Transformation' in this case the government/private sector, although unintentionally (Scoones, Leach and Newell, 2015:177).

The SPMA case underlines the relationship between a citizen-led narrative and a state-led narrative to aid and accelerate a Green Transformation, providing a win-win situation for all actors. The bottom-up approach and citizen-led narrative has influenced the state-led narrative, due to its responsiveness to the pressures for social inclusion. Laws have facilitated access to financing cooperatives and providing education as well as working spaces for the segregation of materials. This is highlighted in Diadema, where the municipality adopted a law that authorises contracts with the cooperatives to pay for their services as part of the municipal recycling programme. Through this, locations have been set up where residents can take recyclables, and waste picker cooperatives are being paid higher incomes (Samson, 2009). It is assumed that the acknowledgement of the benefits of *catadores* and the substantial cost savings for the city (Dias and Alves, 2008) enables these laws to be willingly implemented. These two cases present the impact of leaving an important political actor behind, highlighting the role of politics in achieving transformational change.

Governance and capabilities

The two cases exemplify how the collaboration and relationship between the national and local levels can influence the transition to circular systems. Although Brazil and India both recognise IWP in national policies, at the local level the SPMA integrates these policies more concretely than in Delhi. This can be attributed to the type of approach – decentralised, centralised or a hybrid of both – towards waste management and the inclusion of IWP (Chaturvedi, Arora and Saluja, 2015). According to the analysis, Delhi experiences a more decentralised approach. This means that there is reduced dependency regarding local government and more dependency on citizen and NGO engagement driving the inclusion of IWP. In this type of approach, organisations form alliances with local government and community organisations, which was the case with NGO Chintan Environmental Research and Action Group in partnership with IWP and the NDMC municipality. On the other hand, the analysis indicates that the SPMA utilises a hybrid approach, incorporating aspects of the decentralised and centralised approaches. In this type of approach, certain stages of the waste management chain are completed in both a decentralised (typically collection and segregation) and a centralised (usually recycling and recovering materials) manner. The government of Brazil has implemented various inclusion programmes as well as the National Policy for Solid Waste, which advocates the inclusion of IWP. These initiatives incentivise cooperation between the national and local level as municipalities that incorporate their integrated waste management plans in accordance with this policy obtain federal resources (Besen et al., 2014). The theory of this hybrid model suggests that it overcomes conflict between the formal and informal sectors (Chaturvedi, Arora and Saluja, 2015), implied to be true in the case of SPMA due to the collaboration of models.

It is clear that the inclusion process has differing challenges at different stages of development for nations and cities with differing levels of solid waste modernisation. In both case study cities, more needs to be done to strengthen IWP to organise into waste picker collectives, and the collaboration between different actors is crucial to ensure inclusion. The case of SPMA highlights the impact of collaboration with the support of NGOs and social-governmental relations. However, there is still a gap between policies and legislations for the inclusion of IWP at the municipal and state levels. This also reinforces the notion that there is still a lack of preparedness of governments within marginalised groups in the waste sector, although this is more apparent in India. Going forward, it is vital that policy measures do not further marginalise IWP and consider that livelihoods are central to creating inclusive regulation. To some degree there are still open questions that need to be addressed, specifically in regard to making the CE more inclusive. A deeper understanding of IWP participation during the adaptation of modern waste infrastructure must be closely analysed. It is important to note that the narratives that have allowed IWP to recycle may not be applicable in future scenarios as the development trajectories of India and Brazil could vary greatly in the future.

It is evident that for a Green Transformation to occur, various narratives, drivers and actors are favourable towards IWP and their role in the CE. This can be seen through the SPMA case, where three narratives were dominant: market-led, citizen-led and state-led as well as the collaboration between public, private and civil actors and the cooperation at the national and local levels. However, in Delhi, a market-led and citizen-led narrative is utilised in the inclusion of IWP, presenting little collaboration in closing the WMC through recycling. These findings are in line with the literature on the politics of Green Transformations (Scoones, Leach and Newell, 2015) and the theory that the combination of bottom-up and top-down narratives and approaches as well as multi-level governance shows better outcomes (Bulkeley and Newell,

2010), as experienced in the SPMA case. The findings in the analysis and discussion support the inclusion of IWP and give a comprehensive understanding of the conditions needed for IWP to successfully recycle and become part of the solid waste management in developing cities and bring about the transition to an inclusive circular economy.

Notes

1. It must be noted that this chapter will not discuss the formalisation of IWP through the CE.
2. Taking place across five cities and involving 763 WP (427 women and 336 men).
3. The data in the table was processed by Gupta (2012) with data from Gunsilius, Chaturvedi and Scheinberg (2011) and Scheinberg (2010a)
4. Progressively, there has been evidence of CE transition seen through the micro-level, comprising the private sector, as well as on a macro-level through policies. However, most of these are prevalent in more developed countries (excluding the CE national implementation in China).
5. Attributed to increasing populations (equivalent to 17.84 per cent of the total world population) (Worldmeters, 2016), coupled with a Gross Domestic Product (GDP) growth of 7.6 per cent (Worldbank.org, 2015)
6. Emphasising the recognition and enhancement of the informal sector in collecting and recycling.
7. *Rede Catasampa* serves as the local São Paulo subdivision of the MNCR and comprises 15 recycling cooperatives (Marello and Helwege, 2014).
8. This includes: (1) the Interministerial Committee of Social and Economic Inclusion of Waste Pickers (*Comitê Interministerial de Inclusão Social e Econômica dos Catadores de Materiais Recicláveis*/CIISC) (2003), (2) the National Solid Waste Policy, the national project Cataforte, which was supported by the National Secretary of Solidarity Economy and NGOs (*Secretaria Nacional de Economia Solidária*/SENAES) (Van Zeeland, 2014), and Recycling Bonis Law in 2011 (Dias and Cardoso, 2017).

References

Agarwal, R., Gupta, S.K., Sarkar, P. and Ayushman (2002) *Recycling Responsibility: traditional systems and new challenges of urban solid waste in India*. A Srishti Report.
Besen, G., Ribero, H., Gunther, W. and Jacobi, P. (2014) Selective waste collection in the São Paulo metropolitan region: impacts of the national solid waste policy. *Ambiente & Sociedade*, 17 (3).
BRASIL (2011) Plano Nacional de Resíduos Sólidos: versão preliminar para consulta pública [in Portuguese], September 2011. [Online] Available at: www.mma.gov.br/estruturas/253/_publicacao/253_publicacao02022012041757.pdf.
BRASIL (2013) Diagnosis for the management of urban solid residues [in Portuguese]. National System of Information about Sewage SNIS. Brazil: Ministries of Cities.
Bulkeley, H. and Newell, P. (2010) *Governing Climate Change*. London and New York: Routledge.
Chaturvedi, B. and Gidwani, V. (2011) The right to waste: informal sector recyclers and struggles for social justice in post-reform urban India. In Ahmed, W., Kundu, A. and Peet, R. (eds.), *India's New Economic Policy: a critical analysis*, 125–153. New York: Routledge.
Chaturvedi, A., Arora, R. and Saluja, S. (2015) Private sector and waste management in Delhi: a political economy perspective. *IDS Bulletin*, 46 (3) 7–16.

Chaturvedi, A., Vijayalakshmi, K. and Nijhawan, S. (2015) *Scenarios of Waste and Resource Management: for cities in India and elsewhere*. IDS Evidence Report, 114. Brighton: Institute of Development Studies.

Chi, X., Streicher-Porte, M., Wang, M.Y. and Reuter, M.A. (2011) Informal electronic waste recycling: a sector review with special focus on China. *Waste Management*, 31 (4), 731–742.

Chintan Environmental Research and Action Group (2009) Cooling agents: an examination of the role of the informal recycling sector in mitigating climate change. [Online] Available at: www.chintan-india.org/documents/research_and_reports/chintan_report_cooling_agents.pdf

Consonni, S. (2013) Brazil's incoming e-waste recycling regulations explained. [Online] Available at: https://waste-management-world.com/a/brazils-incoming-e-waste-recycling-regulationsexplained [Accessed 15 June 2015].

Dagnino, R. (2006) Social technology: retaking a debate [in Portuguese]. *Espacios (Caracas)* 27, 1–18. Caracas, Venezuela.

De Brito, D. (2012) God is my alarm clock: WIEGO workers' lives. *Women in Informal Employment Globalizing and Organizing (WIEGO)*. [Online] Available at: https://www.wiego.org/publications/god-my-alarm-clock-brazilian-waste-pickers-story-1

Dias, S. (2006) Waste & citizenship forum: achievements and limitations. In: *Solid Waste, Health and the Millennium Development Goals*. CWG-WASH Workshop Proceedings1–6 February 2006. Kolkata, India.

Dias, S. (2009) Trajectories and memories of the waste and citizenship forums: unique experiments of social justice and participatory governance. Ph.D. thesis. State University of Minas Gerais, Brazil [in Portuguese].

Dias, S. (2011) Statistics on waste pickers in Brazil. WIEGO Statistical Brief No 2. *Woman in Informal Employment Globalizing and Organizing (WIEGO)*. [Online] Available at: http://www.wiego.org/publications/statistics-waste-pickers-brazil

Dias, S.M. and Alves, F.C.G. (2008) *Integration of the Informal Recycling Sector in Solid Waste Management in Brazil*. Study prepared for GTZ's sector project on the promotion of concepts for pro-poor and environmentally friendly closed-loop approaches in solid waste management. Eschborn: Deutsche Gesellschaft für Technische Zusammenarbeit.

Dias, S. and Samson, M. (2016) *Informal Economy Monitoring Study Sector Report: waste pickers*. Cambridge, MA and Manchester: WIEGO. [Online] Available at http://wiego.org/sites/wiego.org/files/publications/files/Dias-Samson-IEMS-Waste-Picker-Sector-Report.pdf.

Dias, S. and Cardoso, S.V. (2017) Negotiating the recycling bonus law: waste pickers & collective bargaining in Minas Gerais, Brazil. In Eaton, A.E., Schurman, S.J. and Chen, M.A. (Eds.), *Informal Workers and Collective Action: a global perspective*. New York: Cornell University Press.

DPCC (2015) Annual review report of DPCC w.r.t MSW for the year 2015–2016. Delhi Pollution Control Committee, Government of Delhi. [Online] Available at: www.dpcc.delhigovt.nic.in/MSW_report2015-2016.pdf

European Commission (2017) Circular economy strategy – environment – European Commission. [Online] Available at: http://ec.europa.eu/environment/circulareconomy/index_en.htm [Accessed 7 June 2017].

Gerdes, P. and Gunsilius, E. (2010) *The Waste Experts: enabling conditions for informal sector integration in solid waste management lessons learned from Brazil, Egypt and India*. Eschborn: GTZ.

Gidwani, V. and Reddy, R. (2011) The afterlives of 'waste': notes from India for a minor history of capitalist surplus. *Antipode*, 43 (5), 1625–1658.

Gill, K. (2010) *Of Poverty and Plastic*. New Delhi: Oxford University Press.

Gunsilius, E., Chaturvedi, B. and Scheinberg, A. (2011) *The Economics of the Informal Sector in Solid Waste Management*. Eschborn: CWG – Collaborative Working Group on Solid

Waste Management in Low- and Middle-income Countries and GIZ – Deutsche Gesellschaft für Internationale Zusammenarbeit (GIZ) GmbH. [Online] Available at: https://www.giz.de/en/downloads/giz2011-cwg-booklet-economicaspects.pdf.

Gunsilius, E., Spies, S., Garcia-Cortes, S., Medina, M., Dias, S., Scheinberg, A., Sabry, W., Abdel-Hady, N., Florisbela dos Santos, A.L. and Ruiz, S. (2011) *Recovering Resources, Creating Opportunities: integrating the informal sector into solid waste management*. Eschborn: GIZ.

Gupta, S. (2012) Waste: the challenges facing developing countries Integrating the informal sector for improved waste management. [Online] Available at: http://www.proparco.fr/jahia/webdav/site/proparco/shared/PORTAILS/Secteur_prive_developpement/PDF/SPD15/SPD15_Sanjay_k_Gupta_uk.pdf [Accessed 25 July 2016].

Gutberlet, J. (2008) *Recovering Resources: recycling citizenship*. Farnham: Ashgate Publishing.

IBGE (Brazilian Institute of Geography and Statistics) (2010) *National Survey of Basic Sanitation* [in Portuguese]. Rio de Janeiro: IBGE.

IPEA (Institute of Applied Economic Research) (2010) *Research About Payment by Urban Environmental Services for the Management of Solid Residues* [in Portuguese]. Brazil: IPEA.

IPEA (Institute of Applied Economic Research) (2012) Diagnosis of solid waste reverse logistics: compulsory research report [in Portuguese]. [Online] Available from: www.ipea.gov.br [Accessed 12 July 2016].

Katusiimeh, M., Burger, K. and Mol, A. (2013) Informal waste collection and its co-existence with the formal waste sector: the case of Kampala, Uganda. *Habitat International*, 38, 1–9.

Köberlein, M. (2003) *Living from Waste*. Saarbrücken: Verlag für Entwicklungspolitik.

Lines, K. and Garside, B. (2014) *Innovations for Inclusivity in India's Informal e-Waste Markets*. Briefing. London: International Institute for Environment and Development.

Lines, K., Garside, B., Sinha, S. and Fedorenko, I. (2016) Clean and inclusive? Recycling e-waste in China and India. Issue Paper. London: International Institute for Environment and Development.

Marello, M. and Helwege, A. (2014) Solid waste management and social inclusion of waste pickers: opportunities and challenges. GEGI Working Paper, 7. Global Economic Governance Initiative. Boston, MA: Boston University.

Medina, M. (2008) *The Informal Recycling Sector in Developing Countries: organizing waste pickers to enhance their impact*. Public Private Infrastructure Advisory Facility. Washington, DC: The World Bank.

Moreno-Sanches, R. and Maldonado, J. (2006) Surviving from garbage: the role of informal waste-pickers in a dynamic model of solid-waste management in developing countries. *Environment and Development Economics*, 11, 371–391.

Nzeadibe, T.C. (2009) Solid waste reforms and informal recycling in Enugu urban area, Nigeria. *Habitat International*, 33, 93–99.

PNRS (2012) *National Plan of Solid Residues, Proposal, Version after Audience and Public Consultation for the National Councils* [in Portuguese]. Environment Ministry and Federal Government. [Online] Available at: www.mma.gov.br [Accessed 15 June 2016] .

Pradhan, J. and Kumar, S. (2014) Informal e-waste recycling: environmental risk assessment of heavy metal contamination in Mandoli industrial area, Delhi, India. *Environmental Science and Pollution Research*, 21 (13), 7913–7928.

Riberio, H., Jacobi, P.R., Besen, G.R., Gunther, W.M.R., Demajorovic, J. and Viveiros, M. (2009) *Selective Pick with Social Inclusion: cooperativism and sustainability* [in Portuguese]. São Paulo: Annablume. [Online] Available at: www.scielo.br/scielo.php?script=sci_nlinks&ref=000158&pid=S1414-753X201400030001500034&lng=pt

Rogerson, M. C. (2001) The waste sector and informal entrepreneurship in developing world cities. *Urban Forum*, 12 (2), 247–259.

Roth, M. (2014) Trash is Cash: waste pickers & economies. *Borgen*. [Online] Available at: www.borgenmagazine.com/trash-cash-waste-pickers-promote-economies/ [Accessed 21 July 2016].

Rutkowski, J.E. and Lianza, S. (2004) Sustainability of solidary enterprises: what role is expected from technology? In Lassance, J.R. (ed.), *Social Technology: a strategy for development* [in Portuguese], 167–186. Rio de Janerio: Bank of Brazil Foundation.

Rutkowski, J. and Rutkowski, E. (2015) Expanding worldwide urban solid waste recycling: the Brazilian social technology in waste pickers inclusion. *Waste Management and Research*, 33 (12), 1084–1093.

Samson, M. (2009) *Refusing to be Cast Aside: waste pickers organising around the world*. Cambridge, UK: Women in Informal Employment: Globalizing and Organizing.

Scheinberg, A., Simpson, M. and Gupta, Y. (2010) *Economic Aspects of the Informal Sector in Solid Waste Management*. Eschborn: GTZ and CWG.

Scheinberg, A. (2010) *Solid Waste Management in the World's Cities: water and sanitation in the world's cities*. London: Earthscan.

Schindler, S., Demaria, F. and Pandit, S.B. (2012) Delhi's waste conflict. *Economic and Political Weekly*, 47 (41), 18–21.

Scoones, I., Leach, M. and Newell, P. (2015) *The Politics of Green Transformations*. London and New York: Routledge.

SNDT Women's University & Chintan Environment Research and Action Group (2008) *Recycling Livelihoods: integration of the informal recycling sector in solid waste management in India*. Study prepared for GTZ's sector project Promotion of Concepts for Pro-Poor and Environmentally Friendly Closed-Loop Approaches in Solid Waste Management. Eschborn: GTZ.

Sthiannopkao, S. and Wong, M.H. (2013) Handling e-waste in developed and developing countries: initiatives, practices and consequences. *Science of the Total Environment*, 463–464, 1147–1153.

Taylan, V., Dahiya, R.P. and Sreekrishnan, T.R. (2008) State of municipal solid waste management in Delhi, the capital of India. *Waste Management*, 28, 1276–1287.

UNCTAD (2017) *Handbook of Statistics 2017 – Population. Fact sheet #11: total and urban population*. Geneva: United Nations Conference on Trade and Development.

Van Zeeland, A. (2014) *The Interaction Between Popular Economy, Social Movements and Public Policies: a case study of the waste pickers' movement*. United Nations Research Institute for Social Development. Paper No 11. Geneva: UNRISD.

Wheeler, K. and Glucksmann, M. (2015) *Household Recycling and Consumption Work*. London: Palgrave Macmillan.

Wilson, D., Velis, C. and Cheeseman, C. (2006) Role of informal sector recycling in waste management in developing countries. *Habitat International*, 30, 797–808.

World Economic Forum (2014) *Towards the Circular Economy: accelerating the scale-up across global supply chains*. Switzerland. [Online] Available at: http://www3.weforum.org/docs/WEF_ENV_TowardsCircularEconomy_Report_2014.pdf [Accessed 25 July 2016].

5

THE ROLE OF WOMEN IN UPCYCLING INITIATIVES IN JAKARTA, INDONESIA

A case for the circular economy in a developing country

Priliantina Bebasari

Introduction

Indonesia is the second biggest marine plastic polluter in the world (Jambeck et al., 2015). Every year the country discharges 200,000 tonnes of plastic into the ocean, or 14.2 per cent of the global total, mainly from Java and Sumatra islands (Lebreton et al., 2017). The critical issue of plastic and waste has attracted the attention of civil society organisations (CSOs) and green businesses since 2010, when the Greeneration Indonesia in Bandung, West Java, initiated the first one-year *Gerakan Diet Kantong Plastik/GDKP* (Plastic Bag Diet Movement) campaign. In 2013, several CSOs like Change.org, Ciliwung Institute, Earth Hour Indonesia, Greeneration Indonesia, Leaf Plus, Indorelawan, and SiDalang joined forces and initiated a national campaign using the same name, GDKP. In December 2015 the Ministry of Environment and Forestry of Indonesia launched a provisional policy stipulating that 'plastic bags are not free'. The policy attempted to reduce plastic consumption by putting a price on plastic bags which were normally given for free to customers at supermarkets. Under this policy, retailers were required to charge customers a minimum price of Rp 200 (around £0.01) per plastic bag in major cities of Jakarta, Tangerang, Bogor, Bandung, Banda Aceh, Balikpapan and Surabaya (Elyda and Agnes, 2016). However, the provisional policy was short lived as the national government failed to issue a stronger legislation to limit plastic bag consumption. In October 2016 the Indonesian Retail Merchants Association (Aprindo) stopped implementing the policy due to an alleged 'lack of legal grounds' (Wright and Waddell, 2017).

Despite the lack of policy on plastic use, waste banks and other green initiatives are mushrooming across the country. Citizens can bring their garbage to waste banks for recycling and receive a payback. This local initiative was first started in

Yogyakarta in 2008 (Lestari, 2012) and has spread to all of Indonesia's 34 provinces with around 5,200 waste banks (Amanda, 2018). Waste banks are a grassroots movement, whereby their potential as social businesses contributing to local economic growth is still to be recognised by the government. Other green initiatives include social enterprises that offer alternative materials to plastic bags as well as waste upcycling. In general, such initiatives engage women workers from the lower income class. The impact of such initiatives on women's resources, agency and achievements (Kabeer, 1999) as well as on waste reduction is yet to be explored.

Despite the growing public awareness of the plastic waste issue, there is still very little information on the circular economy (CE) available in Indonesia. CE research tends to gravitate toward China and Europe, focusing on the environmental dimensions (Merli, Preziosi and Acampora, 2018). Thus, this chapter attempts to fill the literature gap by answering the question **whether a CE contributes to women empowerment and vice versa**. The author hopes that this study, coming from an Indonesian woman's perspective, will initiate a discussion on the CE and its relation to women empowerment in the country.

Circular economy business models and sustainability

The CE is a framework to redesign the economy from a linear take-make/use-dispose model of production and consumption to a closed loop model where the value of products, materials and resources is maintained in the economy for as long as possible (Merli, Preziosi and Acampora, 2018). Two frameworks of analysis are used in this study, namely the ReSOLVE framework – with its six principles of regenerate, share, optimise, loops, virtualise and exchange (Ellen MacArthur Foundation, 2015) – and the typology of CE business models (Table 5.1), to determine whether or not the identified initiatives fall into the CE category.

The STEPS Centre of the University of Sussex and the Institute of Development Studies (IDS) understand the CE in the context of Green Transformations as a social and political process, whereby its sustainability must be linked to the overarching goals of poverty reduction and social justice and to the specific ways that different groups define and refine these goals in particular settings (Leach, Scoones and Stirling, 2007). These multiple pathways to sustainability are interconnected and competing (Leach et al., 2010, referenced in Scoones, Newell and Leach, 2015). A transformation towards sustainability may pose questions that are political and correspond to social justice and institutional change: whether it should be transformational or incremental change, who can use which resources in order to live within the planetary boundaries, who should be involved and with what terms, who wins and who loses, and so on (Scoones, Newell and Leach, 2015). In answering these questions, there are four narratives that provide a different framing of the problem and offer different versions of the solution to sustainability. As a conceptual framework, this study will use the four narratives of Green

TABLE 5.1 Circular business models according to Bocken et al. (2016).

Business model	Description
	Slowing the loops
Access and performance model	Providing the capability or services to satisfy user needs without needing to own physical products
Extending product value	Remanufacturing and refurbishment practices
Classic long-life model	Design of long-life products (durability and repair)
Encourage sufficiency	Prolong product life at the end-user level through durability, upgradability, repair and warranties and a non-consumerist approach to marketing and sales
	Closing the loops
Extending resource value	Transform waste into a valuable resource
Industrial symbiosis	Residual output of an industrial process becomes input for another industry

Transformations as proposed by Scoones, Newell and Leach (2015), which are technology-led, marketised, state-led and citizen-led transformations.

Women empowerment and development

The link between CE and women empowerment can be traced to the demand to connect CE with inequality eradication. CE has been claimed to have links with several Sustainable Development Goals (SDGs), in particular SDG 6 (water and sanitation), SDG 7 (affordable and clean energy), SDG 8 (decent work and economic growth), SDG 12 (responsible consumption and production) and SDG 15 (life on land) (Schröder, Anggraeni and Weber, 2018). The body of literature has also proposed several points of recommendation on how to strengthen CE to tackle inequality in the context of developing economies (Ridpath, Kendal and Gordon, 2017; Gower and Schröder, 2016).

The notion of women empowerment has been emerging as part of the development debate over the last decades. Empowerment is not something that can be bestowed by others, so the space where marginalised people can build agency and make collective action is essential (Rowlands, 1997; Sen, 1997 referenced in Cornwall, 2016:343) to tackle inequality. The literature on women empowerment calls for more than facilitating women's access to assets or creating enabling institutions, laws and policies (Cornwall, 2016:345). Instead, it should be a process of shifting consciousness that locks women into situations of subordination and dependency as well as the engagement with culturally embedded normative beliefs, understanding and ideas about gender, power and change (Cornwall, 2016:345).

Empowerment means a process that enables individuals to maximise opportunities available to them without constraints (Rowlands, 1997). According to

Kabeer (1999), women empowerment is a process by which the marginalised acquire the ability to make strategic life choices which are paramount for people to live the life they want and which can be different from one person to another. Another key aspect of empowerment is that it is a process starting from conditions of disempowerment, or where women's ability to making strategic life choices has been denied. It is only when the failure to realise particular ways of being and doing (functionings), which are valued by people in a given context, is a result of 'some deep-seated constraint on the ability to choose, that it can be taken as a manifestation of disempowerment' (Kabeer 1999:438).

This study uses the three dimensions of empowerment, namely resources, agency and achievement, as proposed by Kabeer (1999), to analyse the selected CE initiatives. The first dimension is resources, which comprises access or actual allocations and future claims to not only material but also social and human resources. An access is reflected in the rules or norms that give certain actors 'authority over others in determining the principles of distribution and exchange' (ibid:437). The second dimension is agency, which is the ability to define one's goals and act upon them. Resources are the pre-conditions and agency is the process to get the outcome. The third dimension is achievements, which are about well-being.

To improve women's well-being through the empowerment process, we must aim beyond making them effective agents in the community and rather support them to be transformatory agents who will challenge the deep roots of gender inequalities (Kabeer, 1999). This is echoed by Sardenberg (2009) in two basic approaches to conceptualising women's empowerment. The first approach is liberal empowerment, which 'regards women's empowerment as an instrument for development priorities' (ibid:5). This is a process where 'individuals engage in to have access to resources so as to achieve outcomes in their self-interest' (ibid:15). Liberal empowerment is similar to women in development, an approach in development that has been criticised for not addressing the issue of power in solving the problem of inequalities. The second approach is liberating empowerment, which perceives women empowerment as 'the process by which women conquer autonomy or self-determination, as well as instrumentally for the eradication of patriarchy' (ibid:5). Sardenberg believes that the main objective of women empowerment is 'to question, destabilise and eventually transform the gender order of patriarchal domination' (ibid:23). Empowerment is a process that 'must address all structures of power' (ibid:13) and 'requires women to first recognise the ideology that legitimises male domination and understand how it perpetuates their oppression' (ibid:11). The liberating empowerment is 'influenced by empowerment as a goal of radical social movements and emphasises the increased material and personal power that comes about when groups of people organise themselves to challenge the status quo through some kind of self-organisation of the group' (ibid:15). Therefore, the literature argues that liberal empowerment is not enough and any empowerment initiatives should aim for liberating empowerment (ibid:15).

Methodology

This study used a qualitative approach to acquire a holistic understanding of the social realities of CE practices (Denscombe, 2010; Marshall and Rossman, 2011) in Indonesia and its relation to women empowerment. I used primary (personal interviews[1]) and secondary data (literature review) for data collection. Data analysis was done using the thematic analysis approach, aiming at identifying, analysing and reporting patterns across data (Flick, 2014). The method is not strictly theory-driven in order to provide a broader explanation of social realities (Denscombe, 2010:304).

The process to answer the main question, whether CE contributes to women empowerment and vice versa, consisted of three steps. The initial step was to identify any initiatives, businesses, projects, programmes or campaigns aiming to reduce waste and plastic pollution in Indonesia (the term initiative refers to all these hereafter), and then to match them against the ReSOLVE framework and Bocken's CE business models (Table 5.1) to see if the initiatives have CE characteristics. The second step was to analyse the initiatives using CE principles and business models as well as how the initiatives have impacted on women workers and local communities. As a last step, the study eventually showed: a) whether the CE initiatives have been promoting a liberal or liberating empowerment, and b) what are the potentials of women empowerment initiatives in supporting CE.

It is worth mentioning that this study has faced challenges in data collection since many initiatives targeting Indonesian women are rather sporadic and short term, and suffer from a lack of records. Therefore, it was not possible to conduct a deeper analysis as to why CE-related initiatives have taken place. This chapter will contribute more to mapping CE initiatives in Indonesia and their relation to women empowerment – and less to evaluating what works and what does not. Recommendations based on academic theories will be provided at the end of this chapter on how to strengthen CE initiatives in Indonesia.

Mapping CE initiatives in the Greater Jakarta Area

The initial search for green initiatives started in Jakarta, the capital city of Indonesia, as it would provide a good case study on CE implementation in a developing country's urban context. Jakarta has an area of 661.52 km^2 with 10.2 million inhabitants and produces 7,147.36 tonnes of garbage per day. However, only 91 per cent of the garbage is managed by the provincial government, leaving 665.61 tonnes of residue (Central Bureau of Statistics, 2016).

During the data collection period, November 2017 to May 2018, 11 CE-related initiatives were identified (Table 5.2). It turned out that Indonesia's Plastic Bag Diet Movement focuses more on campaign and advocacy, while the Waste4Change is a consulting company and *Daur Bunga* is a charity. Evoware, Navakara and Avanieco are green businesses but do not meet any type of CE business models as described by Bocken et al. (2016). These three organisations put more

TABLE 5.2 Mapping of CE-related initiatives identified during data collection.

Green business or initiative	ReSOLVE framework principle	Business model (Bocken, et al.)
Gerakan Indonesia Diet Kantong Plastik (Indonesia's Plastic Bag Diet Movement)	N/A (It is a campaign and policy advocacy)	N/A (It is a campaign and policy advocacy)
Evoware offers seaweed-based packaging	In between regenerate (shifting to renewable energy and material) and exchange (applying new technologies).	N/A (Focus is more on introducing new materials as an alternative to plastic than on discouraging consumptive lifestyles and waste)
Navakara offers non-disposable products such as stainless steel and glass straws as well as bamboo toothbrushes and straws	In between share (prolonging a product's life) by and exchange (replacing conventional materials with advanced, renewable materials)	Classic long life (However, its focus is on introducing new materials as alternatives to plastic, rather than on discouraging consumptive lifestyles)
Avanieco offers products such as bio-cassava bags, ponchos and cups; paper straws and food boxes; wooden cutleries	In between share (prolonging a product's life), regenerate (shifting to renewable energy and material) and exchange (applying new technologies)	N/A (Focus on introducing plant-based materials which are safer for the environment, but not on discouraging consumptive lifestyles)
Waste4change offers consultancy to support corporates, communities and individuals to manage waste responsibly	N/A (No direct implementation, working to support regenerate and loops)	N/A (No direct implementation, working to support industrial symbiosis)
Misu Paper offers note-books and various kinds of papers made from recycled paper	Loops	Extending resource value
Daur Bunga is a charity that collects decorative flowers after events, reuses and donates them to elderly homes.	N/A (Charity initiative using loops principle without a profit-oriented business model)	N/A (Charity initiative to extend resource value, without a profit-oriented business scheme)
Threadapeutic sells bags made from recycled cloth fabric and banners	Loops	Extending resource value and industrial symbiosis
Dreamdelion provided training on upcycling for women, now runs a consultancy agency for non-profit projects	Loops	Extending resource value

Green business or initiative	ReSOLVE framework principle	Business model (Bocken, et al.)
SiDalang provides training on upcycling and business development to local women	Loops and regenerate	Extending resource value
PEKKA is an NGO which supports divorced, widowed and abandoned women to gain access to basic services and economic participation	Loops and regenerate	Extending resource value and industrial symbiosis

focus on introducing new materials that will not harm the environment once a product is disposed, rather than slowing or closing the resource loop. Only five organisations, Threadapeutic, Misu Paper, Dreamdelion, PEKKA and SiDalang, match the criteria of the ReSOLVE framework principles and Bocken's CE business model.

Women's involvement in CE local initiatives

Threadapeutic, Misu Paper, Dreamdelion, PEKKA and SiDalang used household and/or office non-organic waste in their upcycling[2] activities, which included plastic bags, banners, papers, bottles, fabrics, milk cartons, books, tote bags, old clothes and carpets. Besides upcycling, the women groups organised by SiDalang and PEKKA also produced organic fertilisers and created biopore absorption holes filled with organic waste to enhance soil fertility. It was found out that Threadapeutic and Misu Paper did not specifically address women. Hence, to observe any correlation between CE initiatives and women empowerment, this study focuses on three initiatives, namely Dreamdelion, SiDalang and PEKKA, which work closely with women.

Training programmes and marketing

Dreamdelion conducted their CE-related activities in Manggarai slum area in Jakarta, during 2012–2016. The participants were the mothers of children taking part in Dreamdelion's open class. The women came from a low-income class and mostly worked as household cleaners. After four years, Dreamdelion concluded its activities in Manggarai and became a consultant agency.[3] However, its women-focused CE initiative was worth studying. SiDalang started a six-month upcycling training for women in Kampung Dadap (Dadap village), Tangerang[4] in 2013. The first participants were 25 women with an age range of 18–45 years old who were in need of additional income and were struggling to meet their families' basic needs. During 2017–2018, SiDalang expanded its activities to *Rumah Susun*

Tambora (a government-managed public apartment, mostly inhabited by low-income citizens) in Jakarta. SiDalang's creative training programme targeted women (housewives and young women) to provide skills in paper and plastic waste sorting, sewing and making handicrafts. Using presentation and storytelling methods, SiDalang also trained the women to sell handicrafts and become social entrepreneurs. Further, it provided lessons on the 3Rs and how to make biopore holes to increase soil fertility. Following its success in Kampung Dadap and *Rumah Susun Tambora*, SiDalang again expanded its outreach and provided creative/upcycling training to the wider public, with support from local companies (e.g. Mandiri Bank) through their CSR programmes. Both Dreamdelion and SiDalang were initiated by a small group of middle-class educated women.

Differently, PEKKA was founded by Indonesia's National Commission on Stop Violence Against Women (*Komnas Perempuan*) in 2000. The initial objective of PEKKA was to document the lives and to respond to the needs of widowed women in the post-conflict Aceh province (DP3AKB, n.d.). PEKKA operates in 20 out of 34 provinces in Indonesia (MAMPU, n.d.) and organises 2,598 women groups with about 10–100 members per group. Its members are women who have become the head of their households due to the death of their husbands, divorce, or their husbands' inability to support them financially. Single women can become members too when they are the breadwinners, taking care of their parents and siblings. PEKKA does not set any age limit for its members. When a woman is no longer the head of her household, she becomes an 'extraordinary' member. During an interview, it was found that most of the women worked as farm workers and traders. PEKKA's women groups with upcycling enterprises are located in Yogyakarta, Sukabumi, Cianjur, Subang, Tangerang, Karawang and Bekasi.

PEKKA's women groups are not only active in upcycling, but also in waste management advocacy.[5] This has attracted the attention of local governments, who invited the women groups to participate in waste management training, for example the Social Department of Yogyakarta in August 2018 as well as Social Department of Sukabumi and Karawang in 2014. In such training programmes, participants learned various ways to manage waste properly. In PEKKA's upcycling training, the women also learned to create valuable products out of waste such as vases, bags and tablecloths out of plastic waste, and doormats or blankets out of textile waste. The participation in such training as well as a recycling competition organised by the Social Department of Subang[6] have inspired PEKKA's women groups in the area to set up an upcycling enterprise.

Dreamdelion used social media to help the women beneficiaries market the upcycling products, while SiDalang's women groups marketed their own products. In its six-month training programme, SiDalang has included entrepreneurship training that helped women beneficiaries to run their own business independently. PEKKA supported its women groups to market their products using offline channels, i.e. distribution to local communities and markets in the area mentioned.

Challenges, supports from local governments and CSR programmes

These three groups inevitably faced challenges. Dreamdelion and SiDalang reported the lack of human resources as their main challenge. The two initiatives relied on volunteers, who often relied on public transportation, making it difficult to support projects located far away. On the other hand, PEKKA reported a challenge faced by its women groups to market their products since other small upcycling businesses have emerged. Furthermore, the women were preoccupied with advocacy activities in addition to their upcycling business. Due to these challenges, Dreamdelion and PEKKA have ceased offering upcycling business support (though some women groups of PEKKA still produce upcycled products when they receive orders). SiDalang manages to keep providing training on waste management, upcycling and entrepreneurship through its partnerships with the Jakarta government and other local civil society organisations.

Furthermore, SiDalang reported another challenge in obtaining household/office waste for upcycling. In Jakarta, waste has been deemed valuable and many informal workers (waste scavengers) earn their living from collecting and selling waste. Not surprisingly, waste has become inaccessible, with some actors controlling the dumpsites. PEKKA women groups in Sukabumi faced a similar situation following the issuance of Village Law in 2014. The law allows each village to have a village-owned business agency (*Badan Usaha Milik Desa/Bumdes*). Hence, waste materials from local factories that used to be given directly to women groups must be collected and distributed through *Bumdes*. This has increased material and production costs substantially, making the upcycled products more expensive, thus reducing local demand. These cases show that CE and waste management also deal with the issue of power, that is who has access and control over resources.

Despite the challenges, the three CE initiatives have received support from the local governments as well as the private sector in the form of a) invitation to participate in trainings, b) funding from the municipality's social office, or c) funding from the Ministry of Labour, Ministry of Finance or Bank of Indonesia. In most cases, companies offered funding and training through their CSR programmes. It was observed that the funding and training were more event-based rather than long-term partnerships.

Women's empowerment

Except for PEKKA, which is by nature a women-rights organisation, the initiatives did not claim to be women empowerment initiatives. Dreamdelion and SiDalang were established out of concern for waste problems and plastic pollution in the communities in which they operated. The idea to engage local women was based on the founders' personal convictions that women should be offered economic opportunities 'so they can be more productive'. The founders of SiDalang and

Dreamdelion believed that when women were given economic resources, they would invest their money in the education and healthcare of their families, which would eventually improve the well-being of local communities. In contrast, PEKKA aims to address stereotypes and cultural barriers of divorced, widowed and abandoned women, and supported them to get access to basic services and economic participation. PEKKA developed local women groups in various regions and provided them with a space to create collective action. It also introduced social and political empowerment activities such as paralegal training to local women champions (Mathie et al., 2016), who assisted other women in accessing basic services. Therefore, PEKKA women groups are well known in the communities not only for their economic activities, but also for their social activities and political advocacy. However, none of the three initiatives ever conducted a gender analysis to understand the impacts of their women empowerment and CE activities on the communities.

Discussion

Links between CE initiatives, women empowerment and Green Transformation

The three CE initiatives that involve local women in upcycling activities (turning non-organic waste into valuable products) and in business development evidently have implemented the loops principle of the ReSOLVE framework as well as the extending resource value and industrial symbiosis of Bocken's circular business model. Also, PEKKA, SiDalang and Dreamdelion have introduced an element of industrial symbiosis by acquiring waste from factories or offices to be upcycled. The creation of biopore absorption holes using organic waste by the women groups of PEKKA and SiDalang fits very well under the regenerate principle of the ReSOLVE framework.

Using the information collected during the interviews, I analysed the CE initiatives using Kabeer's three dimensions of empowerment, i.e. resources, agency and achievement. The women participating in the upcycling training and social enterprises have shown improved life skills. This is especially true in the projects implemented by Dreamdelion and SiDalang, which engaged with poor women who had never participated in other social projects before. Their participation in the two initiatives has given them options to take part in new economic activities, particularly entrepreneurship – though the interviewees did not provide further details on the women's increased incomes (*resources*).

Through training and coaching, the three initiatives contributed to an improved women's *agency*, making them aware of their identity and value as women as they became more confident to take advantage of new economic opportunities. In particular, SiDalang reported how they have strengthened the women's personal identities. To participate in the training, the women needed to register with their maiden names. It is common in Indonesia for women to be called by the names of their husbands or children (e.g. 'the mother of Siti') after they are married. Thus, the women were surprised to have to register with their own names. This simple requirement has affirmed their identity as a person. They therefore became more confident, creative and experimental during the training. For instance, they started to experiment with mixing other ingredients to make papers, turmeric to produce yellow papers or onion skin to provide texture. Further, the women acquired negotiation skills, which helped them to negotiate with their husbands to allow their involvement in the upcycling activities. In Kampung Dadap, most of the men were fishers and often left home for long periods of time. They expected their wives to stay at home and take care of the family as "all good women do". So, the women would need a permission to be able to participate in the training.

PEKKA engaged with local governments through policy advocacy. It reported that women members in Kampung Sukamaju, Sukabumi who had participated in the upcycling activities have since been asked to participate in drafting their village policy on waste management (following the Village Law enactment in 2014). This was a meaningful *achievement* for the women, who had previously played almost no role in their village governance. Not only that, the women are now actively managing waste banks in their village.

Suen (2013:61) wrote that 'when women are educated, the benefits will be aggregated and magnified and enjoyed by a wider context – the family and the nation'. The upcycling training organised by PEKKA, Dreamdelion and SiDalang have changed the way the women groups perceive waste and its environmental impacts. The women are now aware of the economic value of waste and how upcycling can contribute to reducing environmental degradation. This has helped promote a more sustainable behaviour and lifestyle among the women and within their families.

Evidently, the three CE initiatives have helped create a *space* for women empowerment in Indonesia. However, more research is required to see whether such initiatives can further improve women's access to other economic and social resources which are important for their livelihoods; whether this access can be maintained after a project ends; whether the newly-developed confidence will impact on women's personal goals and ability to act upon these goals; and whether women's achievements, particularly in the case of PEKKA's women groups, are indeed the result of the training programmes on upcycling and business development.

The upcycling initiatives of Dreamdelion, PEKKA and SiDalang indicate that there is a link between sustainability and poverty reduction. According to IBRD (2002), poor people perceive well-being as part of the environment in terms of their livelihoods, health, vulnerability, and empowerment to control their own lives. By tackling the waste issue through upcycling and entrepreneurship, the three initiatives contributed to improving the well-being of poor women and their families. This is in line with the concept of well-being (achievement) as part of empowerment efforts, as suggested by Kabeer (1999).

Analysing the types of Green Transformation (Scoones, Newell and Leach, 2015), the three CE initiatives fit really well under the *citizen-led narratives* since they were established and led by individuals (citizens) concerned with social and environmental problems. In this study, the citizens have initiated and made upskills training and economic resources accessible to people in need. Particularly in PEKKA's case, some groups of women, empowered through the waste management training, have come up with their own ideas to set up an upcycling business. Indeed, the CE initiatives received support from the local government and private sector – in the form of events and training – but it was only after the citizens had taken the first initiative and rolled out some activities.

By now it is clear that the three CE initiatives have approached the issue of women empowerment through *liberal* empowerment and not through *liberating* empowerment, as described by Sardenberg (2009). There are four reasons that may explain this. Firstly, the three CE initiatives did not address the issue of power in their interventions. They focused more on providing women with business skills rather than the skills and resources to improve their position in society or to challenge the roots of inequalities.

Secondly, during the interviews, the CE initiatives seemed to believe that justice is necessarily delivered through a Green Transformation towards CE. This might be due to the fact that green initiatives as well as CE initiatives in Indonesia (see Table 5.2) were, in general, founded by citizens from the middle or high-income classes with a high-level education. No initiatives have been founded by poor women themselves. Therefore, the three CE initiatives have not yet addressed issues such as how they can uphold or improve the position or value of poor women in their relationships with other, more powerful women; how they can ensure that the initiatives are not merely top-down interventions by the middle-class which could harm people at the grassroot level; and how they can ensure that the initiatives do not add to the double burdens (productive and unpaid care work) that women generally bear. It is worth mentioning that during interviews it was discovered that some of the CE enthusiasts seem to unconsciously hold onto gender stereotypes, ranging from 'the women are unproductive because they are seen to roam around when their husbands are at work', to 'women should take up needle work because it is a "women's job"'. Answering these questions mentioned and overcoming these stereotypes might help the three initiatives challenge the root causes of inequalities that keep poor women and their families under poverty.

Thirdly, the three CE initiatives stood alone and did not establish links with other initiatives that can help improve poor women's social and political status. Fourthly, the networking with government was done through events or training rather than policy advocacy, although advocacy usually results in more sustainable or long-term impacts.

CE initiatives and institutional governance

One of the three CE initiatives, SiDalang, reported that women participating in their training have improved their self-confidence. The challenge is to maintain and improve that self-confidence over time and turn it into action, such as the ability of women beneficiaries to be proactive and more entrepreneurial. Cornwall (2016) argued that empowerment is a journey, therefore it is mandatory to move beyond a short-term intervention to long-term oriented activities.

Continuous support from the government would mean creating enabling institutions, laws and policies that assist citizens' CE initiatives to grow. The support can take the form of a regulation mainstreaming the CE in the current economic system, or a specific taskforce to support CE initiatives coming from the citizens as well as to mainstream CE principles within the government itself. Clearly, the CE initiatives cannot thrive on only six-month training sessions. Such initiatives need to be linked to other activities that open an access for women to acquire control of resources and fully participate in decision-making, as advocated by Sardenberg (2009). Marginalised women need long-term support, as opposed to event-based or short-term support, in order to overcome barriers towards an improved well-being. There are social norms surrounding women's participation in the economy (e.g. women should stay at home and do unpaid work) that need to be addressed in order to create an equal society. To achieve this, training and business development alone are not sufficient.

Except for PEKKA, the reviewed CE initiatives do not consider themselves as women empowerment initiatives because they want to be seen as inclusive instead of women-only organisations. The founders feared that their organisations would be labelled as feminist organisations by the public, since there is still a wide-spread negative perception surrounding the feminist movement in Indonesia (Diani, 2016). In addition, the staff of the three initiatives still perceived women empowerment as a separate issue rather than a foundation for a developed society. This is not surprising since many women-oriented organisations operate without knowledge and understanding of power relations or a commitment to fight gender inequality (Desai, 2005). Therefore, it is essential to establish a systematic way to hold such organisations accountable and to ensure they do not perpetuate any stereotypes or gender norms that marginalise and discriminate women. The accountability of organisations to their women beneficiaries is an important part of the empowerment being realised (Kilby, 2011).

PEKKA: a case study of a CE initiative and women empowerment in Indonesia

To answer the question about how women empowerment initiatives can contribute to a circular economy, we can take PEKKA as a case study. PEKKA is the only organisation that links upcycling initiatives to women's participation in decision-making and local governance. Empowered through capacity building, the women groups decided to start their own upcycling business. Although the sale of upcycled products has ceased and the women only serve orders for internal events, they have gained recognition in their communities. They are now involved in local policymaking and their community's waste bank. The upcycling initiative of PEKKA's women groups, alongside PEKKA's advocacy activities, has made the formal power-holders in the community acknowledge these women's role in the community's decision-making. As reported by a PEKKA staff member, some other local upcycling businesses[7] were established following the activities of the PEKKA women groups. This shows how the upcycling ideas and discussions around waste management in the community continue.

In the long term, I believe that women's participation will benefit the economy and ecosystem, particularly through their roles in CE and waste management. With support from the government (through policies) and CE initiatives as well as from a wider community that values women's role in the society, the women can pass the upcycling and entrepreneurial skills on to the next generations. Also, there are other circular business models and principles that can be further explored in Indonesia, such as the slowing-the-loops models (access and performance model, extended product value, classic long-life model, and sufficiency principle), as well as the optimise and virtualise models. To be able to do so, we need women's participation to go beyond upcycling and being duty-bearers at home. We need to see women as active citizens who can make a change in society.

Concluding remarks and recommendations

The cases of Dreamdelion, SiDalang and PEKKA have shown that the circular economy has a potential to bring about women's economic and social empowerment and that educating women on CE principles can help mainstream CE. The three CE initiatives are citizen-led and often lack the resources to undertake long-term interventions with more sustained impacts. Any CE initiatives engaging women also need to be held accountable for their impacts on the women's lives. This can be done by assessing the impacts to the empowerment (economic, social and political) of the women beneficiaries. This study shows that it is strategic to link CE and women's empowerment in order to create greater impacts on poverty reduction and environmental protection.

Several recommendations can be drawn from the three case studies to upscale the impact of CE initiatives and women's participation in Indonesia and beyond:

- The government should create stronger legal, institutional and budget frameworks to mainstream the circular economy.
- Governments and international donors need to develop long-term partnerships with CE practitioners.
- A space for knowledge sharing among CE practitioners as well as between CE practitioners with wider social movements activists needs to be created.
- CE practitioners should explore other CE models and principles.

- CE practitioners and government should link CE to wider efforts to tackle poverty and inequality in the country.
- CE practitioners should assess how CE interventions will impact on women's livelihoods (beyond salary) and agency or ability to make goals and act upon them (beyond confidence).
- CE activities should address social norms and unequal power that put women and marginalised groups under poverty. One way to do so is by linking CE activities with awareness-raising of the women's economic rights.

This study provides an initial assessment of major CE initiatives in Indonesia. Further research is recommended to explore what works (and what does not) in relation to CE and women empowerment. It is also useful to research the potentials of waste banks that are currently mushrooming across the country, and the potentials of social and behavioural change communication *in* CE and *of* CE in other poor and marginalised regions of Indonesia, including the coastal and international border areas.

Notes

1 I used social media (Facebook, Instagram) to reach out to CE-related initiatives which engaged with women since Indonesian citizens and organisations are avid users of social media. I am indebted to Threadapeutic, Dreamdelion, SiDalang and PEKKA, as well as Daur Bunga representatives for their participation in the interviews.
2 'Upcycle' is to upgrade the utility of used or wasted products by introducing artistic elements into the new products.
3 Dreamdelion is now a consultant agency for companies seeking to implement CSR activities with communities. Detailed information such as the number and characteristics of the training participants is not available.
4 Tangerang is part of Jakarta Greater Area, which also includes cities of Bogor, Depok and Bekasi.
5 PEKKA's main interventions include micro-financing for micro and small enterprises, legal capacity building, formal and informal education, political rights and status strengthening, and public health rights (DP3AKB, n.d.).
6 PEKKA could not provide detailed information. In general, non-profit organisations in Indonesia do not keep detailed records.
7 No information can be provided about the other for-profit upcycling businesses. PEKKA does not keep detailed records.

References

Amanda, G. (2018) *Indonesia Kenalkan Bank Sampah ke Dunia Internasional* [in Bahasa Indonesia]. [Online] Available at: www.republika.co.id/berita/nasional/lingkungan-hidup-dan-hutan/18/04/13/p74f88423-indonesia-kenalkan-bank-sampah-ke-dunia-internasional [Accessed 4 August 2018].

Avani Eco (2018) Who we are. [Online] Available at: www.avanieco.com/about-us/ [Accessed 2 May 2018].

Bocken, N.M.P., de Pauw, I., Bakker, C. and van der Grinten, B. (2016) Product design and business model strategies for a circular economy. *Journal of Industrial and*

Production Engineering, 33 (5), 308–320. [Online] Available at: https://doi.org/10.1080/21681015.2016.1172124

Central Bureau of Statistics (Badan Pusat Statistik) Republic of Indonesia (2016) *Jakarta dalam Angka (Statistics of DKI Jakarta Province)* [in Bahasa Indonesia]. Jakarta: BPS Provinsi DKI Jakarta.

Cornwall, A. (2016) Women's empowerment: what works? *Journal of International Development*, 28, 342–359. [Online] Available at: https://doi.org/10.1002/jid.3210

Denscombe, M. (2010) *The Good Research Guide*. Fourth edition. Maidenhead: Open University Press.

Desai, V. (2005) NGOs, gender mainstreaming, and urban poor communities in Mumbai. In: Porter, F. and Sweetman, C. (eds.) *Mainstreaming Gender in Development: a critical view*, pp. 90–98. Oxford: Oxfam GB.

Diani, H. (2016) The F-word: the rise and fall of feminist movements in Indonesia. *Magdalene*. [Online] Available at: www.magdalene.co/news-949-the-fword-the-rise-and-fall-of-feminist-movements-in-indonesia.html [Accessed 6 September 2018].

Dinas Pemberdayaan Perempuan Perlindungan Anak dan Keluarga Berencana (DP3AKB) Provinsi Jawa Barat (2017) *Pemberdayaan Perempuan Kepala Keluarga* [in Bahasa Indonesia]. [Online] Available at: http://bp3akb.jabarprov.go.id/pemberdayaan-perempuan-kepala-keluarga-pekka/ [Accessed 22 August 2018].

Ellen MacArthur Foundation (2015) *Growth Within: a circular economy vision for a competitive Europe*. [Online] Available at: www.ellenmacarthurfoundation.org/publications/growth-within-a-circular-economy-vision-for-a-competitive-europe [Accessed 26 November 2017].

Elyda, C. and Agnes, A. (2016) Minimum plastic bag tax set a negligible Rp 200. *The Jakarta Post*. [Online] Available at: www.thejakartapost.com/news/2016/02/22/minimum-plastic-bag-tax-set-negligible-rp-200.html [Accessed 1 April 2018].

Evoware (2018) Evoware product: seaweed-based packaging. [Online] Available at: www.evoware.id//product/ebp [Accessed 2 May 2018].

Flick, U. (2014) *An Introduction to Qualitative Research*. Fifth edition. London: SAGE Publications, Ltd.

Gerakan Indonesia Diet Kantong Plastik (2018) Tentang Kami [in Bahasa Indonesia]. [Online] Available at: http://dietkantongplastik.info/tentang-kami/ [Accessed 2 May 2018].

Gower, R. and Schröder, P. (2016) *Virtuous Circle: how the circular economy can create jobs and save lives in low and middle-income countries*. Teddington: Tearfund.

IBRD (2002) *Linking Poverty Reduction and Environmental Management: policy challenges and opportunities*. [Online] Available at: https://ec.europa.eu/europeaid/sites/devco/files/publication-linking-poverty-reduction-and-environmental-management-full-report-200207_en.pdf [Accessed 6 September 2018].

Jambeck, J.R., Geyer, R., Wilcox, C., Siegler, T.R., Perryman, M., Andrady, A., Narayan, R. and Law, K.L. (2015) Plastic waste inputs from land into the ocean. *Science*, 347 (6223), 768–771. [Online] Available at: https://doi.org/10.1126/science.1260352

Kabeer, N. (1999) Resources, agency, achievements: reflections on the measurement of women's empowerment. *Development and Change*, 30, 435–464.

Kilby, P. (2011) *The NGOs in India: the challenges of women's empowerment and accountability*. Oxon: Routledge.

Leach, M., Scoones, I. and Stirling, A. (2007) *Pathways to Sustainability: an overview of the STEPS Centre approach*. STEPS Approach Paper. Brighton: STEPS Centre.

Lebreton, L.C.M., van der Zwet, J., Damsteeg, J.-W., Boyan, S., Andrady, A. and Reisser, J. (2017) River plastic emission to the world's oceans. *Nature Communications*, 8 (15611). [Online] Available at: https://doi.org/10.1038/ncomms15611

Lestari, S. (2012) *Bank Sampah, ubah sampah jadi uang* [*in Bahasa Indonesia*]. [Online] Available at: www.bbc.com/indonesia/majalah/2012/07/120710_trashbank [Accessed 4 August 2018].

MAMPU – Australia-Indonesia Partnership for Gender Equality and Women's Empowerment (2018) *PEKKA (Pemberdayaan Perempuan Kepala Keluarga)* [*in Bahasa Indonesia*]. [Online] Available at: http://mampu.or.id/uncategorized/pekka-pemberdayaan-perempuan-kepala-keluarga/ [Accessed 22 August 2018].

Marshall, C. and Rossman, G.B. (2011) *Designing Qualitative Research*. Fifth edition. Thousand Oaks, CA: SAGE Publications, Inc.

Mathie, A., Alma, E., Ansorena, A., Basnet, J., Ghore, Y., Jarrín, S., Landry, J., Lee, N., von Lieres, B., Miller, V., de Montis, M., Nakazwe, S., Pal, S., Peters, B., Riyawala, R., Schreiber, V., Shariff, M., Tefera, A. and Zulminarni, N. (2016) Grass-roots pathways for challenging social and political inequality. In: *World Social Science Report 2016, Challenging Inequalities: pathways to a just world*. Paris: UNESCO Publishing.

Merli, R., Preziosi, M. and Acampora, A. (2018) How do scholars approach the circular economy? A systematic literature review. *Journal of Cleaner Production*, 178, 703–722. [Online] Available at: https://doi.org/10.1016/j.jclepro.2017.12.112

Navakara (2018) About us. [Online] Available at: www.navakara.com/about [Accessed 2 May 2018].

Rowlands, J. (1997) *Questioning Empowerment: working with women in Honduras*. Oxford: Oxfam.

Ridpath, B., Kendal, J. and Gordon, R. (2017) *Going Full Circle: tackling resource reduction and inequality*. Teddington: Tearfund.

Sardenberg, C. (2009) *Liberal Versus Liberating Empowerment: conceptualising empowerment from a Latin American feminist perspective*. Pathways Working Paper 3. Brighton: Institute of Development Studies.

Schröder, P., Anggraeni, K. and Weber, U. (2018) The relevance of circular economy practices to the Sustainable Development Goals. *Journal of Industrial Ecology*. [Online] Available at: https://doi.org/10.1111/jiec.12732

Scoones, I., Newell, P. and Leach, M. (2015) *The Politics of Green Transformations: pathways to sustainability*. Brighton: Routledge.

Suen, S. (2013) The education of women as a tool in development: challenging the African maxim. *Hydra Interdisciplinary Journal of Social Sciences*, 1 (2), 60–76. [Online] Available at: http://journals.ed.ac.uk/hydra/article/view/720 [Accessed 7 September 2018].

Waste4Change (2018) Services. [Online] Available at: http://waste4change.com/# [Accessed 2 May 2018].

Wright, T. and Waddell, S. (2017) How can Indonesia win against plastic pollution? *The Conversation*. [Online] Available at: https://theconversation.com/how-can-indonesia-win-against-plastic-pollution-80966 [Accessed 1 April 2018].

PART III
Policy frameworks and green industrial development approaches

6

THE ARGENTINEAN ZERO WASTE FRAMEWORK

Implementation gaps and over-sight of reusable menstrual management technologies[1]

Jacqueline Gaybor and Henry Chavez

Introduction

In the last two decades, a number of cities around the world, such as Adelaide (2004), San Francisco (2002) and Vancouver (2008), have embraced zero waste principles in order to build a more sustainable society (Zaman, 2015). One of the first and most common actions among policy makers has been to implement legal frameworks and to establish targets regarding the reduction of waste sent to landfills and incinerators, recycling activities and consumer behavioural changes. Given the novelty of the zero waste approach and the variety of experiences around the world, these policies and programmes have been conceived and applied differently, depending on the policy makers' interests and framings (Leach, Scoones and Stirling, 2007).

This global trend has not yet grown firm roots in Latin America. Therefore, the city of Buenos Aires is a pioneer in undertaking the challenge. This city passed the Zero Waste Law in 2005 and implemented it in 2007 (Legislatura de la ciudad autónoma de Buenos Aires, 2005). This law set a number of goals to ban the landfilling of recyclable and compostable waste by 2020. However, a general evaluation of the implementation of this framework done at the end of 2016 suggests that the city is still far from reaching these objectives. What are the possible causes of such a failure? Are they related to the implementation process or do they come from the very design of the policies? How do the policy makers, consumers and other stakeholders interpret these goals and their failure? What have their roles and responsibilities been in this process? Are there any possibilities for improvement?

This chapter tries to answer these questions by critically looking beyond state efforts focused on meeting the goals set in the zero waste framework and exploring the tensions that exist between the fulfilment of these goals and the different framings held by other actors who propose alternative solutions. We will focus specifically on disposable menstrual waste management and the oversight of available reusable solutions. We argue that reusable solutions have been missed by policy makers as clear examples of policy making and implementation failures, due to the limitations in accounting for ambiguities coming from multiple perspectives (Scoones, 2016). In this case in particular, these limitations seem to be reinforced or explained by the concealment of the topic of menstruation, miscommunication, a lack of information and the absence of public debate on these subjects.

Zero waste perspectives: a literature review

Zero waste is an emerging perspective within the scope of waste management and life cycle studies and practices. By redesigning the entire life cycle of every good and service within the economic system, this approach aims to create a 'circular economy' (Winans, Kendall and Deng, 2017) where the output of every production and consumption process becomes the input of another one, eliminating waste and creating economic synergies.

Different definitions of zero waste have been developed. From the idea of sending nothing to landfill or incineration to waste prevention, resource conservation and value recovery, zero waste has been conceived as a systemic approach seeking to redesign products for reuse, repair or recycling back into nature or the marketplace (SF Environment, 2017; Phillips et al., 2011; Zaman, 2015; ZWIA, 2009).

Despite the general agreement on this broad definition, some principles are still debated. For some, zero waste is 'an "end-of-pipe" solution that maximises recycling and minimises waste' (ZWNZT, cited by Tennant-Wood, 2003; MFE-NZ, 2010), but others reject any association with this approach to recycling. For the latter, recycling promotes garbage creation while zero waste is about avoiding waste production by its very design (Palmer, 2005; ZWI, 2017). Nonetheless, everyone seems to agree that applying the 'cradle-to-cradle' principle to the design and manufacturing phases is the best way to ensure the recovery of all resources and to avoid waste.

Likewise, promoting sustainable consumption is expected to minimise waste and environmental damage (Jackson, 2005; Orecchini, 2007; Zaman, 2015). However, studies on consumption and waste generation seem to be scarce among the zero waste literature and very few of them focus specifically on the problem of sustainable consumption and its determinants (Pierre, 2001; Mason et al., 2003; Connett, 2007; d'Arras, 2008; Ball et al., 2009; Orecchini, 2007).

Consumer behaviour and behaviour change depends on individual mindsets, views and attitudes around household goods, energy habits, purchase, the use of domestic appliances and transport behaviour, etc. (Zaman and Lehmann, 2011b). As Jackson (2005) argues, individual behaviour is embedded in social and institutional contexts; therefore, lifestyle changes need a concerted strategy that facilitates more sustainable behaviour by building supportive communities, promoting inclusive societies, providing meaningful work and encouraging purposeful lives. This requires additional efforts from public power.

Cecere, Mancinelli and Mazzanti (2014) suggest that in a European context, waste reduction is seldom associated with socially oriented behaviour or peer pressure, but is reliant on purely 'altruistic' attitudes. However, other studies argue that fees for collection services can also encourage waste reduction (Scheinberg, 2011). Moreover, González-Torre and Adenso-Díaz (2005) showed that recycling habits seem to be related to the type of separated waste, frequency of collection/drop-off, and the distance from selective collection bins. Other studies done in developing countries bring up the importance of culturally specific variables (Chandrappa and Das, 2012; Ekere, Mugisha and Drake, 2009; Sujauddin, Huda and Hoque, 2008; Tang, Chen and Luo, 2011; Zhuang et al., 2008).

Zaman (2014) identifies five key socio-cultural indicators related to zero waste consumption behaviour: household purchase capacity, household expenditures, food consumption, resource consumption and consumption expenditures. According to Zaman, raising awareness and running educational programmes to trigger behaviour change are becoming increasingly important (Zaman and Lehmann, 2011b). From a policy perspective, Davidson (2011) identifies five behavioural instruments that can help to inform and educate consumers: waste audits, school programmes, advertising, training and competitions. Changes in consumption behaviour and product design seem to be more effective in reducing the environmental impact, but their cost and effort can be higher (Stave, 2008).

Timlett and Williams (2011) advocate a more holistic approach considering the 'shared responsibility' among producers, consumers and institutions and its relations with the infrastructure, services and behaviours. This approach is shared by Zaman and Lehmann (2011a), who argue that a zero waste city needs to apply legislation not only on zero landfill and incineration or 100 per cent recycling and recovery of resources, but also on sustainable consumption behaviour and extended producer and consumer responsibility (product stewardship). Guerrero, Maas and Hogland (2013) enlarge this idea of extended responsibility to include also the local authorities. In the same direction, though from a different perspective, a number of scholars advocate increasing education and behaviour change programmes for households, but shifting the focus from recycling to waste prevention and reuse (Barr et al., 2013; Bartl, 2014; Cole et al., 2014; Zacho and Mosgaard, 2016).

Despite the efforts and progress made by zero waste programs implemented around the world, the goal of zero landfills and a 100 per cent resource recovery seems still unreachable. Sanitary waste coming from hospitals, bathrooms, diapers and disposable menstrual management technologies (DMMT) is one among several obstacles to reaching this ambitious target. In a study conducted in the state of São Paulo, Brazil, it was found that 9 per cent of the total solid waste corresponds to this type of waste which is destined for the landfill, while the rest (subdivided into 23 other items) could potentially be reused and recycled (Donnini Mancini et al., 2007). In the city of Buenos Aires, this figure reaches 5.2 per cent of the total solid waste and, as in the Brazilian case, does not have post-use treatments, like for instance plastics or organic waste, and is buried in landfill (FIUBA and CEAMSE, 2016). According to some of the literature, the sanitary risk associated with this kind of waste is high and makes end-of-pipe solutions more complex (Chaerul,

Tanaka and Shekdar, 2008). Whereas for Gerba et al. (2011), absorbent hygiene products in particular do not seem to contribute significantly to the overall pathogen loading present in municipal solid waste landfills.

As we have shown, zero waste literature on consumption and waste prevention is still scarce and the studies on menstrual management are even rarer. Bharadwaj and Patkar (2004), for instance, made an analysis of menstrual hygiene management in Iran, India, Uganda, Kenya and Bangladesh. They criticise the 'hygienist' imaginary constructed around menstruation that has led women to 'hide their blood and throw it away as garbage'. For them, a sustainable solution should start by addressing this ideological problem.

This lack of attention to menstrual hygiene management has also been highlighted by Sommer, Kjellén and Pensulo (2013) in their analysis of menstrual beliefs and behaviours and their relation with existing sanitation systems in low-income countries. Disposable menstrual waste is a big challenge for urban solid waste management. Incinerators seem to be the most common end for this kind of waste in some developing countries, which is not a solution from a zero waste perspective. In their analysis of menstrual management and low-cost menstrual pads, Pathak and Pradhan (2016) propose menstrual cups as an alternative, but they do not develop their analysis further.

Very little work has been carried out with regard to the degradation of sanitary pads and tampons. One study that addresses the topic was conducted by Williams and Simmons (1996) and revealed that buried samples of sanitary pads were shown to maintain their original appearance, and 'showed the greatest tensile strength retention, dropping no lower than 90 per cent' (1996:70). Furthermore, they concluded that 'probably, the longevity of such plastics is a major reason for their abundance and widespread distribution both on river banks and beaches' (ibid:63). Concerns in this regard have been raised by the same FemCare[2] industry as well.

Finally, like Bharadwaj and Patkar (2004), Pathak and Pradhan (2016) point out that the taboo surrounding the topic of menstruation has influenced the lack of research and monitoring of environmental effects vis-à-vis the use of DMMT. As for consumer and waste prevention behaviour, we argue that the solution demands an extraordinary effort in transforming the traditional mindset through education and information programs (Pathak and Pradhan, 2016).

Methodology

This qualitative study is based on a critical analysis of the existing literature and field research conducted in the city of Buenos Aires, Argentina, from August 2016 to January 2017. A total of 28 semi-structured interviews were complemented with participant and non-participant observation. Interviews were conducted with different stakeholders including officials of the city's government, zero waste activists, workers from a recycling cooperative, representatives of NGOs and local producers of reusable menstrual pads and the reusable menstrual cup. The criteria used to select these NGOs were their environmental orientation and their focus on promoting and monitoring the compliance of the 1854 Law. Periodic visits were made to the recycling cooperative Bella Flor to gain a deeper understanding of the functioning of the waste cycle and have a greater dimension with respect to the amount and quality of DMMT waste from people who directly work with this type of waste. This was complemented with participation in events organised on this subject, such as 'The new management of solid waste international workshop' organised by the Ministry of Environment, which gave a broad idea of the current state of affairs and what is envisaged for the management of solid waste at the national level.

The implementation gaps of the Argentinean Zero Waste Law

In 2007, the city of Buenos Aires implemented the 1854 Law for the Integral Management of Urban Solid Waste. The law adopts the zero waste principle to organise the entire life cycle of solid waste. The law initially prohibited waste incineration[3] and set goals and milestones to decrease the amount of waste buried in landfills, thereby establishing concrete targets and deadlines. The first was a reduction of 30 per cent by the year 2010, followed by a reduction of 50 per cent by 2012 and eventually a reduction of 75 per cent by 2017. These milestones were based on the volume of solid waste buried in the 2004 in the Ecological Coordination, Metropolitan Area and State Society (CEAMSE), i.e. 1.49 million tonnes. Another relevant provision of the law consisted of banning landfilling of recyclable and compostable waste by 2020.

It is certain that the three goals were not met. According to CEAMSE (2017), in 2015 a total of 1.15 million tonnes of waste was buried in landfill, which indicates a reduction of barely 23 per cent of the 2004 baseline. After eight years of implementation of the law, the government was still far behind the first target of a 30 per cent reduction, which was supposed to be accomplished by 2010. Under these conditions, it was impossible to reach the final goal of 75 per cent reduction planned for 2017.

The zero waste framework considers the reduction of waste generation as an effective strategy to meet the targets. Thus, the legal instrument forces the Ministry of Environment of the Government of the City to promote awareness generation campaigns to increase society's environmental knowledge and change attitudes toward consumption habits and reuse. However, although reducing waste generation has been awarded a high priority in the law and in its regulations, according to zero waste activists and environmental NGOs' representatives, it has not been of great relevance in the ministry's action plans.

This was confirmed by the ministry officials, who explained that the ministry has mainly focused on two main issues: awareness campaigns to teach the population to sort waste at source and the implementation of various transfer stations and treatment plant projects, including the investment recently made in building a mechanical biological treatment plant (MBT). Concerns about the MBT plant have been raised by environmental NGOs in Argentina (Greenpeace, 2015). One of our interviewees, a senior waste management expert and representative of an international environmental NGO based in Argentina, explained that criticisms of the plant emerged because of the high cost of its construction and ultimately because these investments could lessen the motivation for the population to separate the waste in the household – an action which is free of cost (Personal interview, November 24, 2016).

Zero waste embraces the entire life cycle of solid waste, which starts from the reduction of waste generation by changing consumption habits and reusing. We are clearly talking about measures that go beyond addressing post-consumption sorted and unsorted waste. This is particularly problematic if we consider some of the waste components that do not have a post-use or post-consumption treatment, including waste from DMMT.

Consumption of disposable menstrual management technologies and waste production

It is difficult to estimate the amount of DMMT consumed and its resultant waste. The quantity of disposable pads and tampons used during the menstrual years of a woman will depend on when the menstruation started, the use of contraceptives, the arrival of menopause, pregnancy, illness, the amount of flow, etc. Despite these variations, there are some medical studies that have estimated the amount of DMMT used during the menstrual lifetime of a woman. According to Doctors Shihata and Brody (2014), a woman would use, on average, 17,000 DMMT during her menstrual life. Another study, also coming from the medical literature, makes a more conservative but still worrying estimate. According to Howard et al. (2011), the estimated amount of DMMT used by a woman per year is 169. This is equivalent to 6,000 DMMT used during a woman's menstrual years.

To our knowledge, no similar study has been conducted in Argentina. However, we do have a notion of the quantity of waste generated by the consumption and use of DMMT. The urban solid waste quality study of the city of Buenos Aires (FIUBA and CEAMSE, 2016) probably contains the most up-to-date and detailed information on the composition and quantity of the city's urban solid waste. It is the reference document on urban solid waste for the ministerial officials and NGOs. In the study, both diapers and DMMT are grouped into one single component. Unfortunately, this does not allow precision regarding the exact quantity corresponding to either component.

DMMT and diapers are the fourth highest component in importance, being 5.2 per cent of the total solid waste, preceded by food waste (43.6 per cent), paper and cardboard (14.4 per cent) and plastics (12.6 per cent). To put these numbers in perspective, in 2015 diapers and DMMT waste amounted to a higher volume than pruning and garden waste (4.9 per cent), construction and demolition debris (4.7 per cent) and textile materials waste (4.6 per cent) (FIUBA and CEAMSE, 2016). According to an environmental ministry official, a significant difference between the types of waste mentioned here and DMMT waste is that all the others are *potentially* recyclable, whereas DMMT will end up buried directly in landfills. If not treated,

DMMT waste is not easy to eliminate. The diverse properties of plastic, such as durability and strength, have led to their widespread usage in the manufacture of DMMT, because 'plastics are expected to retain these properties throughout their service lifetime in order to fulfil their required function' (Williams and Simmons, 1996:67).

Over-sight of reusable menstrual management technologies as a zero waste alternative

In Argentina, several reusable menstrual technological innovations have been developed locally by entrepreneurial women, as well as imported to the country from multiple regions of the world, mainly the UK, US, Germany and, very recently, China. Among them, we find menstrual cups and reusable pads.

An 'innovation wave' of reusable menstrual products has taken place since the beginning of the 21st century in many cities around the globe, including Buenos Aires, where the emergence of reusable menstrual management technologies (RMMT) has been fundamentally linked to the protection of the environment and the promotion of health. The first locally produced menstrual cup was launched on the market in 2013, while reusable pads have been produced in different provinces across the country for more than 15 years. Depending on the product, they can have a lifespan of 2 to 10 years, offering a significant reduction in menstrual waste and a low cost in the long run. However, while for most menstrual cup and reusable pad brands the purchase value is amortised within 10 to 12 months, many girls and women still cannot afford them: 'high up-front costs may constitute a barrier to adoption' (Hoffmann, Adelman and Sebastian, 2014). For example, according to Euromonitor Internacional (2016), disposable menstrual pads are the most used products in Argentina to manage menstruation. As of January 2017, the average price of a package of 24 pads was 2.60 Euros. With regard to tampons, still a niche product in Argentina (Euromonitor Internacional, 2016), a box of eight tampons cost on average 1.20 Euros (fieldwork notes, Gaybor, 2017). These prices, contrasted with the high upfront costs of the reusable products, show a clear difference and limitation for certain segments of consumers. The price of one menstrual cup as of January 2017 ranged from 20 to 35 Euro, while the prices for a basic kit of reusable pads ranged from 22 to 25 Euro (fieldwork notes, Gaybor, 2017). All the producers of RMMT interviewed for this study consider the prices fair for the products they make. However, some of them recognised that the price of RMMT could be a limitation to many women and showed a social concern about making these products more accessible. Thus, instead of selling them, they exchange them for other goods or services.

When we speak about reusable pads, we are not talking about the variety consisting of sewn swatches of cloth that women wore before the boom of DMMT. While the principle of reusing cloth is maintained, reusable pads have specific designs and use new materials (i.e. organic and non-organic cotton, Gore-Tex) that go well beyond the earlier versions of sewn or folded cloth. Pads available in Argentina also differ in capacity (e.g. for a regular or a heavy flow) to better accommodate different needs. The selection of reusable products is currently much wider than it used to be. With regard to the menstrual cup, most of the ones available in Argentina are made of medical grade silicon or thermoplastic elastomer, instead of latex-rubber as it was originally designed by Leona Chalmers (Chalmers, 1937). Depending on the brand, they vary in size, colour and design. A great majority of the menstrual cups available in Argentina have an ISO 9001 certification and some are certified by the National Institute of Industrial Technology (INTI) regarding sanitary aptitude. Other brands also have international certifications (i.e. cruelty-free, vegan) to guarantee their quality and safety. However, in the last year, new brands of menstrual cups imported from China have entered the Argentinean market, many of which do not have these certifications and also do not provide details about the materials from which they are made.

Discussion

The challenge of waste prevention

One of the main weaknesses of the Zero Waste Law implementation has been its focus on end-of-pipe solutions. Even if waste prevention and behavioural change are essential parts of the law and in its regulations, these aspects have been neglected in practice.

According to officials of the city government and representatives of the NGOs, several public educational campaigns have been launched since the law was implemented. The communication strategy used in these campaigns focused mainly on promoting at-source waste separation by educating citizens about the different types of waste and the city's waste collection system. However, the interviewees reported contradictory perceptions of the adequacy and effectiveness of these campaigns. Zero waste activists and the representatives of NGOs qualified the campaigns as occasional and to a great extent confusing, since they use technical terms that make it difficult to convey the message to common citizens. An additional problem that was pointed out by a zero waste activist and the representatives of NGOs was the mismatch between the information transmitted by those campaigns and the real state of affairs with regards to the waste collection system. The respondents pointed out that the lack of segregation containers confuses citizens and can discourage behavioural changes. As per government officials, the waste collection system through containers works in the whole city. We were informed by officials that the city is 100 per cent 'containerised'. Nonetheless, government officials warned us of the low rates of waste sorting at source. From their perspective, this failure is explained by a lack of engagement of citizens in recycling practices.

According to activists and NGOs, very little attention is given to promoting behavioural changes related to consumption habits and waste prevention strategies. Accordingly, very few campaigns focused on recycling household waste. This observation is shared by government officials, who agree that these campaigns have principally focused on waste separation at source and not on responsible consumption and reuse. Not surprisingly, the promotion of RMMT as alternatives to prevent waste has been completely absent from public debates and, therefore, from public campaigns. According to government officials, DMMT waste is not considered a priority component within solid waste management. The lack of concern for DMMT waste in the city is reflected by the fact that there is no discussion or research on this subject to develop waste prevention strategies.

DMMT waste is not on the municipal agenda

For most government officials, it is taken for granted that DMMT waste is a component that will not be treated and will go straight to landfill. The interviewees declared that they are not aware of any waste management treatment that could possibly be implemented for this kind of waste besides incineration, which at the time of the interviews was still forbidden. During the interview process, one senior public official expressed awareness of RMMT and considered them as possibly effective preventing alternatives. However, in spite of recognising the possible contribution of RMMT to the prevention of waste generation, it was indicated that it is currently not a priority for the government.

As per non-officials, opening the debates to discuss the ways that RMMT could possibly contribute to achieving the milestones established by the Zero Waste Law is not part of the government's current agenda. Two main reasons preventing progress were highlighted. The first refers to a society that is strongly influenced by consumerism and attitudes of disposability toward objects in general, which results in the lack of questioning the excessive production and consumption of disposable items. The second refers to the strong societal stigma towards menstruation, which prevents discussion of the topic itself and the questioning of the current dominant practices of menstrual management (Felitti, 2016).

As per observations and interviews held in the recycling cooperative Bella Flor, the lack of attention regarding treatment or prevention of this kind of waste could be counterproductive for the recycling process, as it can damage other waste components (i.e. paper), preventing their recycling. Furthermore, according to workers from the recycling cooperative and as noted during the periodic visits, the problem with DMMT waste is not limited to used DMMT, but also to the defective unused pieces that arrive directly from the factory. Surprisingly, unused defective DMMT are not recycled and goes straight to a landfill.

The culture of concealment around menstruation

Confirming insights from the previous literature on the subject, a key theme that recurs in the stories of the producers of RMMT is that the culture of concealment around menstruation has functioned as a barrier to discussing menstrual management. Menstruation has been mostly described as an event that demands secrecy (Felitti, 2016; Rohatsch, 2013; Strange, 2001) and induces feelings of shame and embarrassment when a woman fails to hide it (Houppert, 2000; Vostral, 2008).

According to our interviewees, the concealment of menstruation is constantly reinforced through culturally prescribed behaviour that can come in the form of – amongst others – the advertisement of DMMT (i.e. euphemisms, blue ink is used to replace menstrual blood) (Pessi, 2010); through educational booklets (i.e. synonyms used to replace the use of the word menstruation); or by sexual education sometimes lacking the topic of menstrual management. Argentinean scholar Karina Felitti (2016) has explored the market of RMMT in Argentina. She describes how in events held by promoters of RMMT in public spaces, people engage and participate in activities and conversations. However, she described that while this is an advance in terms of breaking through the concealment of menstruation, it still seems to be uncomfortable for other women to find events that discuss the topic of menstruation in public spaces. Felitti's observations resemble the experiences shared by our interviewees. However, producers of RMMT make a distinction between 'before' and 'now'. 'Before', it was difficult to talk about the topic even with intimate circles of family and friends. They also recalled receiving negative comments and even insults from people through social media and events when they explained about the burden of DMMT waste. However, when they speak about the 'now', although they say that it is still a challenge, they consider that there is a greater openness to discuss and to listen to different perspectives on menstruation management.

For some producers of RMMT, the involvement of other relevant stakeholders, particularly the state, could help open the discussion about the different dimensions of menstruation and raise awareness of the environmental effects of DMMT waste. According to our interviewees, the lack of government policy on this issue limits a large-scale transformation in the management of menstruation and stymies the compliance with the Zero Waste Law.

Designing for sustainable menstrual management

According to the RMMT producers, their involvement in producing RMMT has been self-motivated by their awareness of the negative effects that the use of DMMT entails for women's health, women's economy and the environment. All respondents agreed that designing RMMT could help provide a solution to the complex problem of waste production. However, producers of RMMT emphasised that RMMT alone is not enough to address the complex problem of menstrual management. Most of them pointed out that in order to make this a true technological solution (Rosner, 2004), other cultural, educational, economic and material aspects around menstruation need to be considered and incorporated. While producers of RMMT admit to having high expectations that designing and offering RMMT will reduce the environmental impact of disposal, in their own assessment of their work they identified many contextual challenges they need to overcome.

One of them is the non-familiarity of RMMT: when the menstrual cup or reusable pads are contrasted against traditional DMMT, they can seem strange and complicated and probably not worth the trouble. People's unwillingness to make that 'leap into the unknown' could lead to RMMT remaining unexplored as an alternative option. Another obstacle that was mentioned during the interviewing process has to do with the fact that DMMT are very well established in the Argentinean society. According to our interviewees, their massive use by women across different generations and social classes, particularly disposable menstrual pads (Euromonitor Internacional, 2016), may have prevented any questioning of their environmental, health and economic impact, not only by the users but also by government authorities. This relates to what one of the zero waste activists mentioned. She pointed out that through the years, DMMT have become the standard products, which has created the illusion that are no other options, and thus the only way to manage menstruation. A third obstacle has to do with more practical aspects. DMMT may seem much easier to handle in public spaces than RMMT. As mentioned by one interviewee who produces reusable menstrual pads,

> Once a DMMT is used, it is discarded and replaced by a new one. End of the story. Reusable products demand a different dynamic of use and care. When a reusable pad is changed in a public space, has to be stored in your purse or backpack until you arrive home, where you can wash and dry it. […] I admit there are some challenges in the use of reusable pads. Imagine a school in Argentina, which is not always the safest space for girls. If a girl has a used reusable pad in her backpack, and this is discovered by a classmate, she might be bullied and most likely will have a bad experience […]. The point is that the main issue is not about the reusable used pad in the backpack, but about the stigma that is around menstruation.
>
> *(Personal interview, 2016)*

RMMT not only demand a different dynamics of use and care that may render them difficult to handle in public spaces (i.e. school), they also require an understanding of menstruation outside of stigma. However, there is a fourth challenge, which has to do with having adequate infrastructural conditions to guarantee the safe use of RMMT. Reusable products need to be kept clean to be used safely. Access to clean water to wash them, and in the case of the menstrual cup to boil it, is a first and foremost condition. This raises many questions with regards to their limitation of use by populations that have no permanent access to this resource. Without the necessary care, these products can become vectors of diseases. The last factor, which despite its importance was least mentioned by our interviewees, relates to the economic accessibility of RMMT: their higher upfront costs, although representing longer use value and less waste, are a strong limitation of access even for interested potential users.

Despite multiple challenges, producers of RMMT explained that during the last five years the popularity of RMMT has increased, particularly within the urban and middle class segment of Argentinean society. Among several factors, caring for the environment and preventative health measures seem to be key determinants in their adoption. From the interviews we conducted, we learned that the health benefits that RMMT seem to provide over other DMMT, which can contain harmful chemicals like glyphosate, are a decisive factor in switching to RMMT. On recurrent occasions, producers of RMMT referred to the study conducted by researchers at the University of La Plata (Marino and Peluso, 2015) in Argentina, in which the glyphosate was detected in DMMT. During fieldwork it was possible to see the great press coverage informing citizens about the findings of this scientific study. According to RMMT producers, this has alerted the population and has motivated them to look for alternatives. Nevertheless, a more detailed examination of the ways to manage menstruation – and more profound reasons of why and how this segment of society engages into these new ways – is a future area of inquiry.

Conclusions

Through the analysis of the zero waste framework implemented by the Buenos Aires municipal government since 2007, this chapter aimed to identify the progress and limitations in reaching the targets established within this framework. The information collected through interviews and official reports indicates that it is very unlikely that the programme will reach the goal of zero urban solid waste buried in landfills by 2020. We suggest that one of the main reasons for this failure is that zero waste policy implementation has been focused on end-of-pipe solutions such as recycling and landfilling, instead of waste prevention and promotion of consumer behaviour changes that the zero waste guiding principles had initially promoted. However, as our literature review has shown, this negligence is not unique to the city of Buenos Aires. The lack of analysis of consumer behaviour changes and waste prevention in the zero waste literature suggests that this is a general trend (Ball et al., 2009; Connett, 2007; d'Arras, 2008; Mason et al., 2003; Orecchini, 2007; Pierre, 2001).

Disposable menstrual waste is a particularly strong example of the problems that occur when attention is not given to prevention and reuse. Indeed, our research in Buenos Aires shows that disposable menstrual waste is the fourth highest component of the total solid waste that ends up in landfill. Currently, neither the waste collection systems nor the post-use treatments are designed to deal with DMMT waste. Moreover, government officials do not foresee any alternative procedure, at least not in the medium term. The availability of locally produced alternative reusable menstrual management technologies that can prevent the production of this type of waste highlights the contradictions and limitations between the conception and implementation of such a framework. Hence, even if waste prevention and behavioural change are essential parts of the Zero Waste Law and its regulations, these aspects have been neglected in practice. Several public educational campaigns have been launched, although the communication strategy used in these campaigns has focused mainly on promoting at-source waste separation, and their adequacy and effectiveness has been questioned. Government officials cite the lack of engagement of citizens in recycling practices as the main cause of the low rates of waste sorting at source. Our findings suggest that the government has undertaken very little efforts to bring about behavioural changes related to consumption habits and waste prevention strategies.

We reflected on the multiple dimensions that allow an over-sight of RMMT as a zero waste alternative. The concealment and enduring cultural taboos around menstruation are factors that influence the poor visibility of ecological alternatives. Bharadwaj and Patkar (2004) have assertively argued that a sustainable solution for menstrual management should start by addressing the ideological problem that has led women to 'hide their blood and throw it away as garbage'. Already in the '90s, journalist Karen Houppert (2000) researched the health risks of DMMT, arguing that health risks persist because menstruation remains an under-discussed, under-investigated and concealed topic. In contemporary Argentina, the concealment around menstruation still prevents open discussions on the subject and has functioned as a barrier to the conversation about the environmental impacts of DMMT and the advantages as well as the disadvantages of RMMT. Therefore, it is not surprising that RMMT, as a non-consumerist, low cost in the long term and an effective environmental alternative to prevent waste, has been completely absent from public debates and public campaigns. The concealment has also had an effect on inquiring and demanding actions of those who produce and sell DMMT that cannot be recycled.

Consumer behavioural change with regards to menstrual management encounters different obstacles, such as the non-familiarity of RMMT, higher upfront costs, and difficulties of use/replacement in public areas, among others. Nonetheless, environmental and health concerns among women, especially from the middle class living in urban areas, seem to have contributed to overcoming certain obstacles and have popularised RMMT. Nevertheless, we need to be careful about how we frame this problem – particularly in terms of not placing the environmental responsibility of DMMT waste on the shoulders of women using these technologies. Furthermore, we must inquire into the available alternatives, other than DMMT, for managing menstruation. How much does a common citizen know about the environmental effects of using DMMT? To what extent can technology be a solution when a strong ideological problem around menstruation persists in the context under investigation? Who gets to suggest solutions? We need to think about RMMT as useful tools that can help us meet the essential needs of menstruating women and accomplish other objectives such as diminishing landfill waste. However, we need to reflect on them, not in a vacuum, but rather as components of a larger cultural, educational and economic system. Fostering an open discussion on the complex topic of menstruation in the Argentinean context and advancing research about the advantages and limitations of RMMT and their contribution in achieving the zero waste milestones could be identified as key steps. Nonetheless, the challenge of a society-wide transformation with regards to a change in consumption habits to reduce waste from the use of DMMT necessarily requires alliances among diverse actors, including the state, social movements and the private sector.

Notes

1 The research work for this chapter was partially supported by the Secretaría Nacional de Educación Superior Ciencia y Tecnología from Ecuador (SENESCYT).

2 O.B. claims to have stopped the production of tampons with applicator, arguing that 'plastic applicators are not easy to dispose because they have a life span of 500 to 1,000 years (O.B., 2017).
3 However, on May 3, 2018 the legislature of Buenos Aires modified the law to allow waste incineration (Parlamentario, 2018).

References

d'Arras, D. (2008) Les déchets, sur la voie de l'économie circulaire. *Annales des Mines - Réalités Industrielles* [in French], Novembre 2008 (4), 42. [Online] Available at: www.ca irn.info/revue-realites-industrielles1-2008-4-page-42.htm [Accessed 28 February 2017].

Ball, P.D., Evans, S., Levers, A. and Ellison, D. (2009) Zero carbon manufacturing facility: towards integrating material, energy, and waste process flows. *Proceedings of the Institution of Mechanical Engineers, Part B: Journal of Engineering Manufacture*, 223 (9), 1085–1096. [Online] Available at: http://sdj.sagepub.com/lookup/10.1243/09544054JEM1357 [Accessed 6 April 2017].

Barr, S., Guilbert, S., Metcalfe, A., Riley, M., Robinson, G.M. and Tudor, T.L. (2013) Beyond recycling: an integrated approach for understanding municipal waste management. *Applied Geography*, 39, 67–77. [Online] Available at: http://linkinghub.elsevier.com/retrieve/pii/S0143622812001312 [Accessed 9 April 2017].

Bartl, A. (2014) Moving from recycling to waste prevention: a review of barriers and enables. *Waste Management & Research*, 32 (9_suppl), 3–18. [Online] Available at: http://journals.sa gepub.com/doi/pdf/10.1177/0734242X14541986 [Accessed 28 February 2017].

Bharadwaj, S. and Patkar, A. (2004) *Menstrual Hygiene and Management in Developing Countries: taking stock*. Mumbai: Junction Social. Social Development Consultants. [Online] Available at: www.ircwash.org/sites/default/files/Bharadwaj-2004-Menstrual.doc [Accessed 28 February 2017].

CEAMSE (2017) Statistics. [Online] Available at: www.ceamse.gov.ar/wp-content/uploads/2016/08/web-2015-.pdf [Accessed 15 March 2017].

Cecere, G., Mancinelli, S. and Mazzanti, M. (2014) Waste prevention and social preferences: the role of intrinsic and extrinsic motivations. *Ecological Economics*, 107, 163–176. [Online] Available at: http://linkinghub.elsevier.com/retrieve/pii/S0921800914002092 [Accessed 9 April 2017].

Chaerul, M., Tanaka, M. and Shekdar, A.V. (2008) Resolving complexities in healthcare waste management: a goal programming approach. *Waste Management & Research*, 26 (3), 217–232. [Online] Available at: http://journals.sagepub.com/doi/10.1177/0734242X07076939 [Accessed 28 February 2017].

Chalmers, L. (1937) Catamenial appliance. Patent US2089113 A. [Online] Available at: www.google.com/patents/US2089113 [Accessed 24 April 2017].

Chandrappa, R. and Das, D.B. (2012) *Solid Waste Management*. Environmental Science and Engineering. Berlin and Heidelberg: Springer Berlin Heidelberg. [Online] Available at: http://link.springer.com/10.1007/978-3-642-28681-0 [Accessed 28 February 2017].

Cole, C., Osmani, M., Quddus, M., Wheatley, A. and Kay, K. (2014) Towards a zero waste strategy for an English local authority. *Resources, Conservation and Recycling*, 89, 64–75. [Online] Available at: http://www.sciencedirect.com/science/article/pii/S092134491400113X [Accessed 27 March 2017].

Connett, P. (2007) *Zero Waste: a key move towards a sustainable society*. Binghamton: American Health Studies Project. [Online] Available at: http://www.americanhealthstudies.org/zerowaste.pdf [Accessed 6 April 2017].

Davidson, G. (2011) *Waste Management Practices: literature review*. Halifax: Dalhousie University – Office of Sustainability. Available at: www.dal.ca/content/dam/dalhousie/pdf/

dept/sustainability/Waste%20Management%20Literature%20Review%20Final%20June%202011%20(1.49%20MB).pdf.

Donnini Mancini, S., Rodrigues Nogueira, A., Akira Kagohara, D., Saide Schwartzman, J.A. and de Mattos, T. (2007) Recycling potential of urban solid waste destined for sanitary landfills: the case of Indaiatuba, SP, Brazil. *Waste Management & Research*, 25 (6), 517–523. [Online] Available at: http://journals.sagepub.com/doi/10.1177/0734242X07082113 [Accessed 28 February 2017].

Ekere, W., Mugisha, J. and Drake, L. (2009) Factors influencing waste separation and utilization among households in the Lake Victoria crescent, Uganda. *Waste Management*, 29 (12), 3047–3051. [Online] Available at: http://linkinghub.elsevier.com/retrieve/pii/S0956053X09003158 [Accessed 8 April 2017].

Euromonitor Internacional (2016) *Sanitary Protection in Argentina*. London: Euromonitor Internacional. [Online] Available at: http://www.euromonitor.com/sanitary-protection-in-argentina/report [Accessed 23 April 2017].

Felitti, K. (2016) Menstrual cycle in the XXI century: between the marketplace, ecology and women's power. *Sexualidad, Salud y Sociedad Revista Latinoamericana*, 22, 175–206. [Online] Available at: http://www.scielo.br/pdf/sess/n22/1984-6487-sess-22-00175.pdf [Accessed 24 April 2017].

FIUBA and CEAMSE (2016) *Estudio calidad de los residuos sólidos urbanos (RSU) de la ciudad autónoma de Buenos Aires* [in Spanish]. Buenos Aires: Coordinación Ecológica Metropolitana (CEAMSE) y Facultad de Ingeniería de la Universidad de Buenos Aires (FIUBA). [Online] Available at: http://www.fi.uba.ar/sites/default/files/Estudio%20calidad%20RSU%20-%20versi%C3%B3n%20web.pdf

Gaybor, J. (2018) Menstrual politics in Argentina and diverse assemblages of care. In C. Bauhardt & W. Harcourt (Eds.), *Feminist Political Ecology and the Economics of Care: In Search of Economic Alternatives* (pp. 230–246). London: Routledge.

Gerba, C.P., Tamimi, A.H., Pettigrew, C., Weisbrod, A.V. and Rajagopalan, V. (2011) Sources of microbial pathogens in municipal solid waste landfills in the United States of America. *Waste Management & Research*, 29 (8), 781–790. [Online] Available at: http://wmr.sagepub.com/cgi/doi/10.1177/0734242X10397968 [Accessed 28 February 2017].

González-Torre, P.L. and Adenso-Díaz, B. (2005) Influence of distance on the motivation and frequency of household recycling. *Waste Management*, 25 (1), 15–23. [Online] Available at: http://linkinghub.elsevier.com/retrieve/pii/S0956053X04001503 [Accessed 8 April 2017].

Greenpeace (2015) *Las Plantas MBT: una falsa solucion para cumplir con la ley de basura cero* [in Spanish]. Greenpeace Argentina. [Online] Available at: http://www.greenpeace.org/argentina/es/informes/Las-Plantas-MTB-una-falsa-solucion-para-cumplir-con-la-ley-de-basura-cero/ [Accessed 16 May 2018].

Guerrero, L.A., Maas, G. and Hogland, W. (2013) Solid waste management challenges for cities in developing countries. *Waste Management*, 33 (1), 220–232. [Online] Available at: http://www.sciencedirect.com/science/article/pii/S0956053X12004205 [Accessed 27 March 2017].

Hoffmann, V., Adelman, S. and Sebastian, A. (2014) Learning by doing something else: experience with alternatives and adoption of a high-barrier menstrual hygiene technology. *Menstrual Hygiene Day*. [Online] Available at: https://observer.american.edu/cas/economics/news/upload/Hoffman-Paper.pdf [Accessed 24 April 2017].

Houppert, K. (2000) *The Curse: confronting the last unmentionable taboo: menstruation*. First edition. New York: Farrar, Straus and Giroux.

Howard, C., Rose, C.L., Trouton, K., Stamm, H., Marentette, D., Kirkpatrick, N., Karalic, S., Fernandez, R. and Paget, J. (2011) FLOW (finding lasting options for women). *Canadian Family Physician*, 57 (6), e208–e215. [Online] Available at: http://www.ncbi.nlm.nih.gov/pmc/articles/PMC3114692/ [Accessed 23 April 2017].

Jackson, T. (2005) *Motivating Sustainable Consumption: a review of evidence on consumer behaviour and behavioural change*. Surrey: Sustainable Development Research Network. [Online] Available at: http://www.sustainablelifestyles.ac.uk/sites/default/files/motivating_sc_final.pdf [Accessed 6 April 2017].

Leach, M., Scoones, I. and Stirling, A. (2007) *Pathways to Sustainability: an overview of the STEPS Centre approach*. STEPS Approach Paper. Brighton: STEPS Centre. [Online] Available at: https://steps-centre.org/publication/pathways-to-sustainability-an-overview-of-the-steps-centre-approach/ [Accessed 20 May 2018].

Legislatura de la ciudad autónoma de Buenos Aires (2005) *Ley de Gestión Integral de Residuos Sólidos Urbanos* [in Spanish]. Buenos Aires: Boletín Oficial de la Ciudad Autónoma de Buenos Aires.

Marino, D. and Peluso, L. (2015) *Residuos de Glifosato y su metabolito AMPA en muestras de algodón y derivados* [in Spanish]. III Congreso de Médicos de Pueblos Fumigados-UBA. La Plata: EMISA - Universidad de La Plata.

Mason, I.G., Brooking, A.K., Oberender, A., Harford, J.M. and Horsley, P.G. (2003) Implementation of a zero waste program at a university campus. *Resources, Conservation and Recycling*, 38 (4), 257–269. [Online] Available at: http://linkinghub.elsevier.com/retrieve/pii/S0921344902001477 [Accessed 6 April 2017].

MFE-NZ (2010) *The New Zealand Waste Strategy: reducing harm, improving efficiency*. Wellington: Ministry for the Environment of New Zealand. [Online] Available at: http://www.mfe.govt.nz/sites/default/files/wastestrategy.pdf.

O.B. (2017) Tampones o.b. – Un tampón que se adelanta a su tiempo [in Spanish]. *Tampones O.B España*. [Online] Available at: http://www.tamponesob.es [Accessed 23 April 2017].

Orecchini, F. (2007) A 'measurable' definition of sustainable development based on closed cycles of resources and its application to energy systems. *Sustainability Science*, 2 (2), 245–252. [Online] Available at: http://link.springer.com/10.1007/s11625-007-0035-8 [Accessed 6 April 2017].

Palmer, P. (2005) *Getting to Zero Waste*. Sebastopol, CA: Purple Sky Press.

Parlamentario (2018) Autorizaron la quema de basura en hornos controlados. *Parlamentario*, 3 May 2018. [Online] Available at: http://www.parlamentario.com/noticia-109420.html [Accessed 16 May 2018].

Pathak, N. and Pradhan, J. (2016) Menstrual management and low-cost sanitary napkins. *Economic & Political Weekly*, 51 (12), 27. [Online] Available at: http://www.indiawaterportal.org/sites/indiawaterportal.org/files/menstrual_management_and_low_cost_sanitary_napkins_epw_2016.pdf [Accessed 28 February 2017].

Pessi, M.S. (2010) Tabú y publicidad: el titular en avisos publicitarios gráficos de productos para el período menstrual [in Spanish]. *Tonos Digital: revista electrónica de estudios filológicos*, 19, 25. [Online] Available at: https://dialnet.unirioja.es/servlet/articulo?codigo=3439315 [Accessed 20 May 2018].

Phillips, P.S., Tudor, T., Bird, H. and Bates, M. (2011) A critical review of a key waste strategy initiative in England: zero waste places projects 2008–2009. *Resources, Conservation and Recycling*, 55 (3), 335–343. [Online] Available at: http://www.sciencedirect.com/science/article/pii/S0921344910002223 [Accessed 27 March 2017].

Pierre, F. (2001) Becoming a zero waste to landfill facility in the USA. In *Proceedings Second International Symposium on Environmentally Conscious Design and Inverse Manufacturing*, pp. 760–765. [Online] Available at: http://ieeexplore.ieee.org/document/992463/?reload=true.

Rohatsch, M. (2013) ¿Estás venida? Experiencias y representaciones sobre la menstruación entre niñas de 12 a 15 años [in Spanish]. *Avatares de la Comunicación y la Cultura*, 6. [Online] Available at: http://ppct.caicyt.gov.ar/index.php/avatares/article/view/2868 [Accessed 24 April 2017].

Rosner, L. (Ed.) (2004) *The Technological Fix: how people use technology to create and solve problems*. First edition. New York: Routledge.

Scheinberg, A. (2011) *Value Added: modes of sustainable recycling in the modernisation of waste management systems*. Ph.D. thesis. Wageningen: Wageningen University.

Scoones, I. (2016) The politics of sustainability and development. *Annual Review of Environment and Resources*, 41 (1), 293–319. [Online] Available at: http://www.annualreviews.org/doi/10.1146/annurev-environ-110615-090039 [Accessed 20 May 2018].

SF Environment (2017) Zero waste. *San Francisco Department of the Environment*. [Online] Available at: https://sfenvironment.org/zero-waste/overview/zero-waste [Accessed 5 April 2017].

Shihata, A. and Brody, S. (2014) An innovative, reusable menstrual cup that enhances the quality of women's lives during menstruation. *British Journal of Medicine and Medical Research*, 4 (19), 3581. [Online] Available at: http://www.journalrepository.org/media/journals/BJMMR_12/2014/Apr/Shihata4192014BJMMR9640_1.pdf [Accessed 23 April 2017].

Sommer, M., Kjellén, M. and Pensulo, C. (2013) Girls' and women's unmet needs for menstrual hygiene management (MHM): the interactions between MHM and sanitation systems in low-income countries. *Journal of Water, Sanitation and Hygiene for Development*, 3 (3), 283. [Online] Available at: http://washdev.iwaponline.com/cgi/doi/10.2166/washdev.2013.101 [Accessed 28 February 2017].

Stave, K.A. (2008) Zero waste by 2030: a system dynamics simulation tool for stakeholder involvement in Los Angeles solid waste planning initiative. In: *Proceedings of the 26th International Conference of the System Dynamics Society*. Athens, pp. 20–24. [Online] Available at: http://www.ewp.rpi.edu/hartford/~ernesto/S2014/SHWPCE/Papers/SW-Prevention-Integration/Stave2008-ZeroWasteby2030.pdf [Accessed 5 April 2017].

Strange, J.M. (2001) The assault on ignorance: teaching menstrual etiquette in England, c. 1920s to 1960s. *Social History of Medicine: the journal of the society for the social history of medicine*, 14 (2), 247–265.

Sujauddin, M., Huda, S.M.S. and Hoque, A.T.M.R. (2008) Household solid waste characteristics and management in Chittagong, Bangladesh. *Waste Management*, 28 (9), 1688–1695. [Online] Available at: http://linkinghub.elsevier.com/retrieve/pii/S0956053X07002255 [Accessed 8 April 2017].

Tang, Z., Chen, X. and Luo, J. (2011) Determining socio-psychological drivers for rural household recycling behavior in developing countries: a case study from Wugan, Hunan, China. *Environment and Behavior*, 43 (6), 848–877. [Online] Available at: http://eab.sagepub.com/cgi/doi/10.1177/0013916510375681 [Accessed 8 April 2017].

Tennant-Wood, R. (2003) Going for zero: a comparative critical analysis of zero waste events in southern New South Wales. *Australasian Journal of Environmental Management*, 10 (1), 46–55. [Online] Available at: http://www.tandfonline.com/doi/abs/10.1080/14486563.2003.10648572 [Accessed 5 April 2017].

Timlett, R. and Williams, I.D. (2011) The ISB model (infrastructure, service, behaviour): a tool for waste practitioners. *Waste Management*, 31 (6), 1381–1392. [Online] Available at: http://www.sciencedirect.com/science/article/pii/S0956053X10006288 [Accessed 6 April 2017].

Vostral, S. (2008) *Under Wraps: a history of menstrual hygiene technology*. Lanham: Lexington Books.

Williams, A.T. and Simmons, S.L. (1996) The degradation of plastic litter in rivers: implications for beaches. *Journal of Coastal Conservation*, 2 (1), 63–72. [Online] Available at: http://www.springerlink.com/index/QM2N33NN46H74647.pdf [Accessed 23 April 2017].

Winans, K., Kendall, A. and Deng, H. (2017) The history and current applications of the circular economy concept. *Renewable and Sustainable Energy Reviews*, 68, 825–833.

[Online] Available at: http://linkinghub.elsevier.com/retrieve/pii/S1364032116306323 [Accessed 28 February 2017].

Zacho, K.O. and Mosgaard, M.A. (2016) Understanding the role of waste prevention in local waste management: a literature review. *Waste Management & Research*, 980–994. [Online] Available at: http://wmr.sagepub.com/content/early/2016/07/11/0734242X16652958.abstract [Accessed 28 February 2017].

Zaman, A.U. (2014) Identification of key assessment indicators of the zero waste management systems. *Ecological Indicators*, 36, 682–693. [Online] Available at: http://www.sciencedirect.com/science/article/pii/S1470160X13003567 [Accessed 27 March 2017].

Zaman, A.U. (2015) A comprehensive review of the development of zero waste management: lessons learned and guidelines. *Journal of Cleaner Production*, 91, 12–25. [Online] Available at: http://linkinghub.elsevier.com/retrieve/pii/S0959652614013018 [Accessed 28 February 2017].

Zaman, A.U. and Lehmann, S. (2011a) Challenges and opportunities in transforming a city into a 'zero waste city'. *Challenges*, 2 (4), 73–93. [Online] Available at: http://www.mdpi.com/2078-1547/2/4/73 [Accessed 6 April 2017].

Zaman, A.U. and Lehmann, S. (2011b) Urban growth and waste management optimization towards 'zero waste city'. *City, Culture and Society*, 2 (4), 177–187. [Online] Available at: http://www.sciencedirect.com/science/article/pii/S1877916611000786 [Accessed 27 March 2017].

Zhuang, Y., Wu, S.-W., Wang, Y.-L., Wu, W.-X. and Chen, Y.-X. (2008) Source separation of household waste: a case study in China. *Waste Management*, 28 (10), 2022–2030. [Online] Available at: http://linkinghub.elsevier.com/retrieve/pii/S0956053X07002619 [Accessed 8 April 2017].

ZWI (2017) The zero waste institute. [Online] Available at: http://zerowasteinstitute.org/ [Accessed 5 April 2017].

ZWIA (2009) ZW definition. *Zero Waste International Alliance*. [Online] Available at: http://zwia.org/standards/zw-definition/ [Accessed 5 April 2017].

7
ASSESSMENT OF THE CIRCULAR ECONOMY TRANSITION READINESS AT A NATIONAL LEVEL

The Colombian case

Claudia Lorena Garcia and Steve Cayzer

Introduction

The current linear economic model, based on the activities of take-make-and-dispose, has been widely recognised as unsustainable and unable to meet contemporary environmental, social and economic needs (Lacy and Rutqvist, 2015). This model has national consequences which are dependent on the income and context of the country (Ellen MacArthur Foundation, 2013; Weizsäcker et al., 2014; Lacy and Rutqvist, 2015).

In this chapter we consider low and middle-income countries, in particular Colombia. Here a high dependency on the mining and agricultural sectors leads to vulnerability in the face of global economic deceleration and price volatility (MarketLine, 2015). Furthermore, environmental challenges are also present in the country. According to the Global Footprint Network, Colombian biocapacity has been progressively decreasing since 1961 (Global Footprint Network, 2016) and the rising volume of waste produced has meant significant challenges in solid waste management (OECD/ECLAC, 2014).

In this context, given the attractiveness of the circular economy (CE) to meet current economic and environmental challenges, global efforts have been undertaken to understand and contextualise this paradigm under different circumstances (UNEP, 2006; Ellen MacArthur Foundation, 2013). However, there is still little literature regarding a CE transition in the context of low and middle-income countries, opening room to research and propose strategies to enable a transition given the specific circumstances of these countries. Thus, this chapter explores the enablers for a CE transition in low and middle-income countries presenting an enabling framework which is the baseline to evaluate the CE transition readiness in Colombia and to suggest some interventions required to achieve a more circular economy in the country.

Circular economy transitions

Despite the efforts to advance a CE concept, the transition towards a CE is still at a very early stage globally and has faced different challenges due to the deep-rooted linear economy structure and mind-set (Haas et al., 2015; Ghisellini, Cialani and Ulgiati, 2016; Lieder and Rashid, 2016). Nevertheless, some efforts to the transition toward circular economies, the most documented cases being in Europe and China, have displayed relevant insights regarding different approaches and challenges to this economic model (Su et al., 2013; Ghisellini, Cialani and Ulgiati, 2016; Lieder and Rashid, 2016). These approaches have been the result of the different understanding of a CE, the specificity of the local contexts and the particular motivations to pursue a CE (European Commission, 2014; Murray, Skene and Haynes, 2015; Preston, 2012; Ghisellini, Cialani and Ulgiati, 2016; Lieder and Rashid, 2016).

In the case of Europe, materials security has been one of the main drivers for adopting a CE model because of the high dependence on imports of raw materials (Preston, 2012). A transition started as a private sector initiative with a focus on financial benefits (Murray, Skene and Haynes, 2015), following a *bottom-up approach* where the actions of the private sector have gradually been supported by policies and governmental institutions (Geels, Elzen and Green, 2004; Ghisellini, Cialani and Ulgiati, 2016). A formal CE strategy has been the 'Resource Efficient Europe' (Preston, 2012), which aims to increase the resource efficiency and waste minimisation through technological innovation as a core activity, in addition to the Circular Economy Package with measures for the whole cycle from manufacturing to waste management with a special focus on product design (European Commission, 2018). Some examples of the practices of a CE in Europe are the zero waste programs, eco-industrial parks and eco-cities as well as some initiatives at the company level such as eco-design and eco-labelling schemes (Ghisellini, Cialani and Ulgiati, 2016). Although some progress in resource productivity has been observed as the result of these efforts, Europe still remains resource dependent (Ellen MacArthur Foundation, 2015b). Thus, more policy support to develop inner-cycle activities (repair, reuse, refurbishment, remanufacturing) and to design more durable products is needed (WWF, 2015; Stahel and Clift, 2016; European Commission, 2014). The importance of supporting sharing schemes and 'product as a service' models in Europe is well documented (Tukker, 2015; Hobson and Lynch, 2016; Ellen MacArthur Foundation, 2015b).

On the other hand, the Chinese approach has shown its urgency in solving the environmental crisis by implementing mostly initiatives rooted in the industrial ecology with a focus on recycling (Su et al., 2013). A *top-down* national political strategy towards a CE has been adopted with the Circular Economy Promotion Law, which intends to balance environmental and social concerns caused by the rapid Chinese economic expansion (Bigano, Sniegocki and Zotti, 2016; Ghisellini, Cialani and Ulgiati, 2016). This top-down approach has been developed with formal rules for companies to comply with whilst they receive governmental and international organisations' support (Geels, Elzen and Green, 2004; UNEP, 2006). The actions enclosed within the CE promotion law have been to some extent effective regarding the reduction of resource consumption and greenhouse gas emissions (Su et al., 2013). However, in practice there is little evidence of industrial strategies such as eco-design (Su et al., 2013), and this approach to CE has been considered less profitable and materially efficient than other approaches, such as those proposed by the performance economy (Stahel and Clift, 2016). Furthermore, there are different challenges in the implementation of all the initiatives, for instance, the lack of incentives to promote them or reliable information to track the progress in circularity, the deficit of advanced technology, and the lack of public awareness to support the undertaken programs (Bigano, Sniegocki and Zotti, 2016; Su et al., 2013).

From the analysis of CE global practices and due to the lack of a comprehensive framework to support a CE transition, a recent work has proposed a 'practical strategy' for CE implementation with a concurrent top-down (national effort) and bottom-up approach (company effort) (Lieder and Rashid, 2016). The reason for this proposal is the inverse motivations among stakeholders (nations, governmental bodies, society and industrial business enterprises) and a strong need to 'align and converge' these motivations to support a CE transition. One example of these inverse drives is the conflict between governmental bodies and enterprises' interests. The former aims to ensure the environmental rights by controlling industrial activities. However, the industry is likely to underestimate its environmental impacts because of the prioritisation of economic benefits and growth in a more competitive environment. As a result, enterprises overlook the benefits from a sustainable strategy such as the CE, and they avoid investing in sustainability practices. This concurrent approach is proposed under the assumptions that technology is sufficiently mature to support a CE transition at large scale and that there is no prioritisation of environmental or economic benefits at the expense of each other in the implementation of this strategy.

Colombia in the context of circular economy

Nowadays, the urgency of shifting towards a more sustainable economic model is relevant not only to the developed world but also in the context of low and middle-income countries (Lacy and Rutqvist, 2015; Ellen MacArthur Foundation, 2013; Weizsäcker et al., 2014). The Colombian context is not an exception to this situation. Firstly, the country's economy has been highly dependent on the extractive sector and a low-performing industry, requiring action to generate more value-added and regenerative activities. Secondly, the population in the region has been growing at a very high rate, leading to urbanisation and considerable waste management problems that are calling for effective solutions (Hoornweg and Bhada-Tata, 2012; OECD/ECLAC, 2014). And thirdly, as a consequence of these two factors, an increasing environmental degradation has been evidenced. Thus, the search for triple bottom line solutions is important for the country.

Regarding the Colombian economic structure, the service sector represents a value-added share of 57.7 per cent, the manufacturing sector accounts for a share of 22.9 per cent, and finally the extractive industry (mining and agriculture) represents a share of 19.4 per cent (OECD/ECLAC, 2014). Colombia has been to some extent dependent on the extraction of commodities, mainly oil, coal, and agricultural products, accounting for 15 per cent of the total GDP and 70 per cent of the total exports in 2014 (MarketLine, 2015). Although the extraction of these commodities has played a significant role in the Colombian economy's growth, it has lately been vulnerable to global economic deceleration, which has led to a reduction of exports to key trade partners such as Europe, the USA and China (MarketLine, 2015). Table 7.1 summarises some of the data regarding the Colombian economy compared to OECD countries. According to this data, a marked dependency on natural resources can be seen in Colombia.

Moreover, the industrial sector has presented a low performance in recent years (MarketLine, 2015), showing a slow growth of 2.6 per cent during 2010–2013 (DNP, 2014) and a reduced capacity to diversify the economy (OECD, 2015). As has been reported by the OECD (2015), only 25 per cent of the manufacturing firms in the country are classified as high technology or medium-high technology companies, leading to a low level of exports of high-tech products. This performance has been partly the result of the low national expenditure in RandD, at just 0.17 per cent of the GDP, compared to other countries in the region such as Brazil, which invests 1.24 per cent of the GDP, and OECD countries' investment of 2.40 per cent of GDP (see Table 7.1). In addition, a low involvement by the private sector in the efforts to foster technological innovation, with a small business expenditure of 0.04 per cent of the GDP, is also evident in the country (MarketLine, 2015; Cornell University, INSEAD and WIPO, 2016).

TABLE 7.1 Current state: economy, recovery of materials and innovation in Colombia.

	COLOMBIA	OECD
Economy	Natural resources rent: 3.6% GDP (2015) Oil, gas, coal & coffee: 70% of total exports (2014) (MarketLine, 2016)	Natural resources rent: 0.5% GDP (OECD, 2017)
Innovation	Country R&D investment: 0.17% of GDP (2014) R&D contribution of business enterprises: 0.04% of GDP (MarketLine, 2016)	Average R&D investment: 2.40% of GDP (2015) R&D contribution of business enterprises: 1.65% of GDP (OECD, 2017)
Recovery of materials at the end of use	Recovery of materials: 1% (OECD, 2014) Informal recycling activities (50,000 families) (National Recycling Association, 2017)	Recovery of materials: 24% high-technology recycling facilities (OECD, 2017)

A second issue to consider is the growing population of the country and its environmental consequences. Colombia, as most of the countries in Latin America and the Caribbean, is following a trend of overpopulation, extended urbanisation and uncontrolled waste generation (Hoornweg and Bhada-Tata, 2012). By 2025, a growth of 20 per cent in the total population is expected with a concentration in urban areas of 80 per cent, leading to an municipal solid waste generation per capita of 1.5 kilogrammes per capita per day (the average for the OECD is 2.2 kilogrammes per capita per day) (Hoornweg and Bhada-Tata, 2012). The growing volume of waste generated and the lack of a proper infrastructure for its treatment has been alarming, leading to urgent calls to develop instruments to reduce waste generation and increase means to incorporate the materials into the value chains (OECD/ECLAC, 2014). Some reports claim that in Colombia, 92 per cent of the municipal solid waste goes to the landfill, 7 per cent goes to non-controlled sites, and just 1 per cent is recovered by composting and recycling with a precarious infrastructure and by waste pickers who are not incorporated into the formal labour sector (OECD/ECLAC, 2014; Wilson et al., 2006). Indeed, recycling by waste pickers is the only activity reported for materials recovery at the products' end-of-use and has become the livelihood of around 50,000 families that have been working in creating cooperatives to gain a 'voice' and formalise their activities (Medina, 2000). This situation contrasts with strategies developed in OECD countries, where recycling rates are higher and high-technology infrastructure is used.

A third aspect to consider is ecological degradation because of the current practices in the country. Although the Colombian ecological footprint did not exceed its total biocapacity by 2012, this last one has consistently been decreasing since 1961 (Global Footprint Network, 2016). As Vallejo, Pérez Rincón and Martinez-Alier (2011) have claimed, this decreasing biocapacity could be the consequence of the growing extraction of natural resources to supply foreign and local demand. The effects of this ecological degradation are multiple and include aspects such as pollution, deforestation and soil degradation (Vallejo, Pérez Rincón and Martinez-Alier, 2011; United Nations Environmental Programme, 2016). Equally significant has been the rise in CO_2 emissions of 30 per cent between 2004 and 2012 (MarketLine, 2015). The rising concentration of carbon dioxide has a direct impact on the country's air quality and has become a public health concern since it is already affecting the most vulnerable population: 1 child among 100,000 died because of outdoor air pollution in 2008 (United Nations Development Programme, 2015; MarketLine, 2015).

Finally, the Colombian public and private sectors, being aware of current environmental and economic challenges, have undertaken different initiatives. Some governmental initiatives have been driven by the ambition to be part of the OECD, including the National Policy on Sustainable Consumption and Production of 2010. This policy aims to change current production and consumption patterns towards sustainable development, and it states the concept of 'closed-loop production' (MADS, 2010). However, the practice of these efforts has been to some extent limited by a lack of coherence between the country's economic plans and its environmental goals. As the report of OECD/ECLAC (2014) claimed, there is a lack of joint efforts between the different ministries that should participate in sustainable development, namely the economic ministries (agriculture, industry, mines and energy) and the environmental ministry, and this is stopping current green initiatives. Moreover, in 2012 the country signed the OECD Declaration on Green Growth (OECD/ECLAC, 2014), which has led to the inclusion of green growth strategies in the National Development Plan 2014–2018 and to the execution during 2017–2018 of the 'Green Growth Mission' which aims to prepare strategies to increase the country's competitiveness and productivity in enhancing natural capital (DNP, 2017). On the other hand, the corporate sector, which is subject to environmental regulations, has been implementing some strategies based on optimising their processes and has been looking for opportunities to develop sustainability strategies which benefit them in terms of profits and competitiveness.

Methodology

Since there is no study on CE transition in Colombia, this research was conducted through an explorative and qualitative approach using interviews for primary data collection. The interview questions were semi-structured and used open-ended questions to obtain a broader view of the participants. These interviews were designed to get relevant information related to the enablers of a CE transition in low and middle-income contexts and the specific circumstances that would support or challenge a CE in Colombia. The interviewees were selected based on their particular experience in different fields, such as innovation, sustainability, waste management and circular economy, and on their involvement in different sectors: academia (AC1), environmental consulting (CON1), non-governmental organisations (NGO1) and a neutral point of view from an expert from the Ellen MacArthur Foundation (EMF1) who has been researching the CE globally. The information from the interviews allowed us to assess the proposed enabling framework, focusing in particular on drivers, barriers, enablers and current policies for a CE in Colombia.

An enabling framework for a CE in low and middle-income countries

As has been pointed out in previous studies, a circular economy transition starts by establishing the right enabling conditions (Garcia Caicedo, 2016). Thus, given the specific circumstances of low and middle-income countries, such as the lack of formal recycling activities (Gower and Schröder, 2016; Egerton-Read, 2016) and some development gaps in terms of technology and infrastructure(Garcia Caicedo, 2016; Egerton-Read, 2016), an enabling framework is proposed that represents the baseline to evaluate the Colombian's readiness to move to a CE and to identify the main interventions that would facilitate this transition (Garcia Caicedo, 2016). This framework was based on the available literature (Lieder and Rashid, 2016; Ellen MacArthur Foundation, 2015b) and the insights from interviewing EMF1 about the aspects that should be considered in the context of low and middle-income countries.

Figure 7.1 presents the proposed enabling framework which has a systemic perspective by converging the interests and responsibilities of all stakeholders (nations, governmental bodies, society and manufacturing industries). According to the suggestions presented by Lieder and Rashid (2016), this framework reflects a concurrent top-down and bottom-up approach towards the same goal: 'an economy which is environmentally and economically regenerative'.

In the top-down part of the framework, *policy and legislation* reinforce any approach to CE (Lieder and Rashid, 2016; Stahel and Clift, 2016). They are materialised by the establishment of a *support infrastructure* (Lieder and Rashid, 2016) that includes a fiscal framework(Lieder and Rashid, 2016; Ellen MacArthur Foundation, 2015c; Stahel and Clift, 2016), education (Ellen MacArthur Foundation, 2015c), financing schemes (Ellen MacArthur Foundation, 2015b), IT infrastructure, and a safe recovery of materials at the end-of-use stage (Garcia Caicedo, 2016). Simultaneously, the bottom-up part of the framework emphasises support of new business models (Lieder and Rashid, 2016; Stahel and Clift, 2016), using design-led approaches to production such as Cradle to Cradle design (Garcia Caicedo, 2016) and the design of reverse supply chains aided by information communications

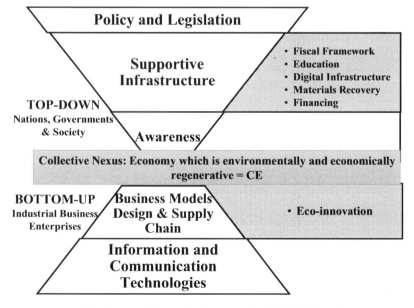

*This framework is based on Lieder and Rashid's (2016) work

FIGURE 7.1 Enabling framework for a CE transition.

technology for remanufacturing, refurbishment, repair, recycling and reuse (Lieder and Rashid, 2016; Ellen MacArthur Foundation, 2015c). Some implications for low and middle-income countries are presented in the following:

Policy and legislation: A CE transition requires the support of policies and regulations, and joint strategies should be present among economic and environmental policies.

Fiscal framework: This includes the adoption of sustainable taxation and a fiscal structure where jobs generation is enabled to include informal groups into the economy through a tax shifting from people to resources. This means reducing the taxes in the hiring process, considering that human capital is a renewable resource (Preston, 2012; Stahel and Clift, 2016), and taxing the extraction of non-renewables and the generation of waste and emissions (Stahel and Clift, 2016). Some of the drivers behind this enabler include the protection of natural and human capital (Stahel and Clift, 2016) and the facilitation of the recovery of materials based on non-renewables since the taxes would make these materials more valuable, promoting their recovery rather than their production (Stahel, 2016).

Education: This is a fundamental enabler to prepare future professionals with a new economic paradigm, and build the skills needed to drive circular innovation. It is suggested that 'circular and systems thinking' be included in school and university curricula (Ellen MacArthur Foundation, 2015c). As it is supported by Stahel (2016), 'a broad "bottom up" movement will emerge only if SMEs can hire graduates who have the economic and technical know-how to change business models'.

Digital infrastructure: The current digital revolution must be embraced to generate new business models that allow the circulation of value, such as product service systems and the integration of more people into the economy through co-creation. However, it is also pointed out that the digital revolution in the linear economy has been used to reduce costs and to extract value by the use of robots and to promote mass production; this can represent a risk for countries which have labour-intensive industries and do not have a clear strategy towards sustainable development (Garcia Caicedo, 2016)

Materials recovery: Within a CE framework, there is a need for a profitable and safe recovery of materials. A circular economy proposes value-adding activities that ensure the materials are neither hazardous nor toxic for the people who handle them (Garcia Caicedo, 2016). Thus, it strongly discourages the way in which informal waste collection and recycling practices are carried out in low and middle-income countries because it represents health risks for those involved (Garcia Caicedo, 2016)

Financing: Economic support is strongly needed for all actors in value chains. As it has been pointed out by the Ellen MacArthur Foundation (2015b), this transition involves costs to invest in new assets, digital infrastructure, RandD, retraining, market penetration of new products and importantly to support the disrupted industries by new business models that can arise.

Social awareness: Raising awareness about CE is key to enabling a CE considering the important role of customers in the economy and the change of mindset to support this shift of economic paradigm.

Business models, product design and supply chain: Industrial business enterprises play an important role to transitioning towards a CE. Business should be rethought to be profitable and to enhance natural capital. In this way, companies should implement circular strategies supported by product design and reverse supply chains; some examples include remanufacturing business models or product services systems.

ICTs in businesses: ICTs enable the development of reverse supply chains by using. For example, the internet of things.

Circular economy transition readiness in Colombia

The proposed enabling framework was applied to the particular Colombian context in order to assess the country's readiness for a CE transition. This assessment was carried out by reviewing the country's policy landscape, some reports on sustainability, innovation and ICTs' performance as well as the insights obtained from primary data through interviewing three experts working on sustainable development in Colombia. The results of this assessment are summarised in Table 7.2 and are expanded in this section.

Overall, from a systemic perspective and according to the suggestion of the convening interests and responsibilities of all stakeholders for a successful circular economy transition, it seems that the country does not have a clear strategy yet. While some governmental institutions, such as the National Planning Department, the Ministry of Commerce, Industry and Tourism and the Ministry of Environment and Sustainable Development, have been promoting circular and green economy programs, other important ministries such as the Ministry of Mines and Energy, the Ministry of Information Technologies and Communications and the Department of Innovation (Colciencias) have not been fully involved on those initiatives. In the private sector, the National Business Association of Colombia has been leading some post-consumer programs to recover hazardous materials at the end of use, aiming to meet collection goals of the current extended producer responsibility regulations; however, these efforts have focused on meeting regulations rather than as a business strategy to grow companies. Considering the important role of consumers in making this transition happen, there is still little incentive for them to become actively involved in recovery activities. Thus, raising awareness and education efforts should be reinforced to foster a citizen-led transition.

Policy, fiscal framework and subsidies

At first, the lack of a unified sustainability vision and different narratives of the economic development plans and the environmental goals of the country can potentially hinder the accomplishment of a CE transition. In fact, there are

TABLE 7.2 CE transition readiness in Colombia.

	Enablers	Assessed implications	Main insights
Top-down: nations, governmental bodies, society	Legislation and policy	National policies supporting a CE	Lack of a joint strategy to support a CE transition Conflicts of interest between the economic and environmental plans/policies Economic dependency on the oil and gas industry
	Supportive infrastructure		
	Fiscal framework and subsides	Tax shift from labour to resources Subsidies Tax incentives	High non-wage costs Low taxes perceived from the extraction of non-renewables compared to international standards Implicit subsidy for fossil fuels extraction and use Subsidies for biofuels, incentives to promote cleaner technology in industry, transportation and energy generation
	Education	Integration of CE and systems-thinking in universities curricula	Lack of understanding of the basic concepts of CE such as systems thinking
	Financing	Financial support for sustainable and CE initiatives	There are no financial schemes to support circular business models
	IT infrastructure	4G and broadband infrastructure ICT access ICT use	Need for an increased quality of the broadband services, larger coverage of the digital network and more ICT appropriation among businesses and society
	Profitable and safe recovery of materials	MSW management Hazardous waste management	Current materials recovery system is informal and it is not safe for people involved in these activities. It is difficult to recover materials Not well-developed markets for recovered materials
	Social awareness	Public communication and information campaigns	Low awareness of circular economy principles among private, public and academic sectors

Bottom-up: industry, businesses, enterprises	CE business models, design-led approach to production & supply chain	Eco-innovation CE business models implementation	Low private investment in R&D activities Lack of technical support for SMEs to appropriate CE business models and design-led approaches to production Lack of collaboration between producers and suppliers through the supply chain to support circular businesses
	Information and communication technology	ICTs use as a mean of creation of new business models	Low appropriation of ICTs to develop new businesses among the private sector

conflicting interests between the National Policy for Sustainable Production and Consumption of 2010 and the National Development Plan for 2014–2018 (OECD/ECLAC, 2014). While the former stresses the need of generating more sustainable consuming and manufacturing patterns among the different sectors (MADS, 2010), the latter supports mining as a key 'locomotive' of economic growth (DNP, 2014). Some positive signs towards achieving joint economic and environmental strategies can be seen in the current Green Growth Mission, which aims to structure the country's roadmap and new policies to enable green growth with the three pillars of water, energy and materials efficiency towards a circular economy, positioning the bio-economy as a competitive sector to diversify the economy and to increase job opportunities to support a green growth strategy (DNP, 2017).

Current policy contradictions are reflected in some subsidies and incentives that promote the extraction and consumption of non-renewables. According to the interview with EMF1, a CE transition is enabled by eliminating subsidies to extract non-renewables. However, in Colombia the transport fuel producers have access to some benefits through the current pricing formula, which has resulted in an implicit subsidy. This formula proposes an increase of 3 per cent of the producer income when this is lower than the export parity price, which has been the trend in latest years (Garcia and Calderon, 2013). Moreover, there are incentives to use diesel in some aquaculture activities and to foster the extraction of oil and coal in the territory to attract foreign direct investment (OECD/ECLAC, 2014). There are some positive sings boosted by the National Policy on Sustainable Production and Consumption, which include subsidies to use biofuels and to implement clean technologies in industry, energy generation and transportation.

Moreover, current fiscal structure could be enhanced by creating a 'pro-circularity framework'. As acknowledged in the interview with EMF1, a key lever is the shift of taxes from labour to resources; in his words: 'move taxes from taxing people to taxing the extraction of non-renewable resources'. This move supports the employment of people and promotes the preservation of non-renewables. However, as has been established by the OECD (2015), the high non-wage costs in Colombia are a limitation for the generation of jobs, with an average of 49 per cent of the total payroll in 2015. Despite some efforts to reduce the payroll tax, the incentives for companies to incorporate the informal sector are low.

In addition, there are some limitations to current environment-related taxes. According to the report of OECD/ECLAC (2014), the largest fraction of these tax revenues comes from transport fuel taxes (energy products) and a smaller fraction from motor vehicle taxes. Although, there are taxes related to the extraction and use of non-renewable resources, there are some limitations in this fiscal system. The first limitation to consider is that there are no taxes for energy products used for stationary purposes, for example, electricity or cooking fuels (OECD/ECLAC, 2014). Furthermore, these taxes do not consider the environmental impacts of fuel use or vehicle performance (OECD/ECLAC, 2014). A final limitation is regarding the royalties obtained by the government from the extraction of fuels since they have been considered low compared to international standards, while the investment of these royalties is ineffective in mitigating the environmental impacts generated by the mining activities (OECD/ECLAC, 2014).

This ambiguity in policies supporting economic growth and sustainability reflects the Colombian economy's dependency on non-renewable resources. which could represent a carbon lock-in that is able to stop any sustainability strategy. As has been pointed out by Skene and Murray (2015), countries with large supplies of fossil fuels have little incentive to develop alternative sources of energy. A circular context would imply abolishing this dependency, thus affecting some industries, which the country must be financially able to support as they adjust to the new paradigm (Ellen MacArthur Foundation, 2015a). In addition, a strong 'regenerative industry' framed within the CE principles should be developed to protect the country's economy. Hence, there is an opportunity to integrate the economic and environmental policies in the country, seeking to generate strategies that reduce this dependency and are aimed to promote other sectors with a high potential for circularity and high value added (Ellen MacArthur Foundation, 2015a). For instance, some successful case studies in developing countries have shown opportunities in the use of organic waste for composting or energy generation and clusters for repair and remanufacturing around automotive factories (Gower and Schröder, 2016). Any further initiative towards circularity in the country should consider these strategies.

Education and awareness raising

Colombia, in its Sustainable Production and Consumption National Policy, has stated a vision towards capacity building in the academic sector to foster a cultural change towards sustainability. Thus, this policy highlights the need for strengthening technical skills for sustainable production and consumption at universities and the National Service of Learning (SENA) as well as the importance of environmental education at the primary school level supported by the Environmental Education Policy of 2001 (MADS, 2010). As has been established by MADS (2010), the goal of this policy is to include permanently relevant modules on sustainable production and consumption in the curricula of the engineering, economics and businesses programmes. This strategy aimed to cover 20 per cent of the educational institutions in 2014 and to achieve a coverage of 40 per cent in 2019 (MADS, 2010). Furthermore, some consumer information campaigns have been adopted to support the Extended Producer Responsibility programme under the name 'Cierra el Ciclo' (close the loop).

However, according to the interview with the environmental consultant, Colombia does not have a system that promotes educational aspects on CE and systems thinking. Some practical exercises in design for sustainability have shown a lack of a complete understanding of the key concepts to support a transition:'(. . .) during some workshops with universities, students did not have a real notion of the meaning of sustainability and they did not think about the surrounding system and the life cycle of a product (. . .) they are not able to have a systemic view in order to design a product'

(CON1). Therefore, it is suggested to integrate fundamental CE concepts within universities' curricula as a basis to train human capital with a different mind-set.

Digital infrastructure

Concerning the country's digital infrastructure, there have been remarkable efforts and strategies to boost this sector, with the prospect of generating digital innovations for export. As a result, the country has extended broadband connection and 4G services, achieving a coverage of 44 per cent and 40.46 per cent, respectively, in 2014 (DNP, 2014). Furthermore, 17 entrepreneurship centres to develop digital innovations have been created and 17,000 SMEs have been given support to adopt ICTs (DNP, 2014).

However, there are some challenges to create a circular economy aided by the digital revolution. The first one is the broadband services quality improvement, which allows the development of third generation digital solutions for a CE, such as the internet of things (OECD, 2014; DNP, 2014). Thus, it is necessary to have higher bandwidths than those provided by the current broadband in Colombia. Therefore, investments to improve the current broadband service must be executed (DNP, 2014).

Secondly, an increased coverage of the digital network is needed, as EMF1 highlights the importance of this expansion to enable more people to innovate:

> (. . .) increase the digital network, because that would allow more people to come out and try to create more things (. . .) this is really exciting and disruptive because it means that people with the entrepreneur spirit can grab a hold of it.
>
> *(EMF1)*

Despite the efforts to improve the IT infrastructure, just 51.7 per cent of the population has access to the internet, 43.6 per cent of the households in urban areas use the internet and just 6.8 per cent do so in the countryside (DNP, 2014). Expanding the infrastructure to difficult access zones will require more investment and incur high costs to due to geographical limitations and the dispersion of the municipalities, which can stop the access to ICTs in the national territory (DNP, 2014).

In addition, it is necessary to increase public interest and awareness among the population about the benefits of ICTs to create business opportunities and increase productivity, because some reports have shown that users prefer applications for entertainment or communication rather than using the technology for the development of new business or productivity applications (DNP, 2014). Finally, even though there is an ecosystem for digital entrepreneurship for the development of applications and services, it is missing financial assistance, especially seed capital, to enable digital innovations (OECD, 2014).

Material recovery

Colombia also presents opportunities to establish strategies to promote a safer and profitable recovery of materials. The country has an acceptable waste collection rate of 80 per cent (OECD/ECLAC, 2014) and since 2011 it has been implementing an extended producer responsibility initiative to manage hazardous waste streams in some products, such as used tires, light bulbs, computers and batteries (OECD, 2016). However, the waste management strategy has been acknowledged as being corrective without avoiding the generation of waste (OECD/ECLAC, 2014), and some gaps have been identified in the existing regulations and initiatives.

The first gap is the lack of formal activities to treat the collected materials. To be more precise, informal recycling is the only registered method for materials recovery and, according to reports, just 1 per cent of the waste is recycled or composted (OECD/ECLAC, 2014). Furthermore, there are concerns regarding the conditions in which the informal sector carries out these activities. As AC1 highlighted, they lack a proper infrastructure to treat the waste and are exposed to different hazards: '(. . .) because of the manual processes involved in recycling activities, there are some health problems such as ergonomics, toxicity, infections, and injuries'. Therefore, it is important to consider how toxic waste streams are because they can represent an imminent risk for people who handle them, but it is more important to implement strategies to avoid the circulation of toxic materials. Here is where the producer's liability plays an important role by designing, producing and commercialising safe products and where legislation must establish an effective regulatory framework to control toxic materials flows.

Apart from this, the markets for recycled products are not well developed, and they do not attract investors because of the high costs they could represent (OECD/ECLAC 2014). Thus, CON1 states that recycling is barely done: '(. . .) recycling depends highly on the market and if there is not demand of the materials or there are not facilities to process them, the recycling activities stop'.

As a consequence, there is a limited scope of extended producer responsibility (EPR) initiatives, because the collected materials are usually stored or exported before being recycled abroad (OECD/ECLAC, 2014). A recent assessment of this initiative has shown that while the collection rates have been improving annually, companies lack the capabilities to recover components from post-consumer waste and there are no targets for recycling rates (OECD, 2016).

In view of these findings, the country presents opportunities to develop value adding businesses by safely and profitably recovering materials in waste streams (Garcia Caicedo, 2016). A suggestion is to look for strategies to incorporate the informal sector in activities that increase and optimise material and cash flows (Egerton-Read, 2016), such as the repair, remanufacturing and cascading of organic wastes (Gower and Schröder, 2016).

Financing

Currently, there are some financing mechanisms for companies to accomplish 'green growth' strategies in Colombia (OECD/ECLAC, 2014); some examples include the Clean Technologies Fund (CTF), which supports sustainable urban transport systems and energy efficiency projects, and the Green Protocol, an agreement between the government and the financial sector to provide financing options for green projects. However, the scope of these initiatives seems to be short, as the report of DNP (2014) has claimed a considerable part of the Colombian business community does not have access to more efficient and cleaner technologies, due to economic and financial barriers as well as a lack of effective loan programmes. Moreover, according to the interview with NGO1, Colombia is lacking financial schemes for circular economy initiatives.

Business models, product design and supply chain and ICTs in businesses

Finally, the assessment of the enablers at the company level has led to some insights. According to the interviews conducted in the country (AC1, CON1, NGO1), there is an opportunity to enable the transition towards a CE by supporting the innovation and the establishment of specific strategies oriented to *eco-innovation* among the industrial sector (Garcia Caicedo, 2016). In this way, CON1 claimed that it is highly important to encourage the largest companies, which already have the capability to invest in RandD, to develop the skills to implement design-led approaches to production, such as Cradle to Cradle, biomimicry, and others (Garcia Caicedo, 2016). However, the country faces challenges related to public and private investment in RandD activities, which is just 0.2 per cent of the GDP, compared to other countries of the region such as Brazil, who invest 1.24 per cent of the GDP and, more industrialised countries like Germany, where the RandD investment reaches 2.84 per cent of the GDP (Cornell University INSEAD and WIPO, 2016).

In addition, the importance of SMEs to facilitate a transition is emphasised by EMF1, citing the example of bioMASON, which is disrupting the construction industry by producing bricks from microorganisms: '(. . .) SMEs are important if they've got the right enabling conditions, this is where change comes from'. However, some projects in closed-loop production in the country have shown that it is necessary to incorporate efficient business support schemes in order to enable SMEs to participate in a CE. These schemes should reinforce companies' technical capabilities, create collaboration platforms between clients and suppliers, and provide financial mechanisms to implement changes in design and production processes. According to the opinions of NGO1 and CON1, these have been the biggest barriers to enabling a CE transition in the SMEs sector:

> (. . .) SMEs do not have strong negotiation leverage over their raw materials suppliers in order to support any change in their processes. Moreover, it is missing leadership to enact and train the companies to make the implementation of these initiatives (CE initiatives) possible.
>
> *(CON1)*

> (. . .) It is missing lines of credit and incentives for circular economy initiatives. This is one of the biggest barriers that are stopping to SMEs to continue with the transition towards a circular economy.
>
> *(NGO1)*

Moreover, it is important to encourage the generation of digital innovations among populations and enterprises, since there is a lack of interest in using ICTs to develop new business models and to use them productively (OECD 2014). To quote EMF1

> (. . .) SMEs can use the digital revolution to outcompete the bigger players, and this is why, in emerging economies, it can be exciting, because you can help to develop or apply technologies in markets which are not yet as fossilized as they are in the more developed world.
>
> *(EMF1)*

For this to happen, it is necessary to implement strategies to increase technological appropriation among the micro-enterprise sector.

A final consideration regarding the efforts oriented towards a CE supported by companies is the need to raise awareness about the economic and environmental benefits among the enterprises, since a lack thereof can be a significant barrier for the transition (Garcia Caicedo, 2016). NGO1 has indicated that one crucial factor to developing CE related projects has been to showcase the benefits of CE among the industrial sector.

Conclusions

The CE stands out as a compelling alternative to the unsustainable linear economy. Despite the progress on CE transition in Europe and China, there has been little research on its feasibility in the context of low and middle-income countries. Therefore, this chapter explored the CE in Colombia, where a high dependency on natural resources has led the country's economy to become vulnerable to the consequences of a global deceleration and price volatility. The CE potentially offers solutions to create a more diverse and competitive industry which can support the SDGs.

Firstly, this explorative study established the enabling conditions that would allow the country to transition towards a more circular economic system. An enabling framework has a systemic perspective and converges the different interests and responsibilities of all stakeholders (nations, governmental bodies, society and manufacturing industries). This framework reflects a concurrent top-down and bottom-up approach to enable a CE transition. The enablers to support this transition in low and middle-income countries include: policy and legislation, a 'pro-circularity' fiscal framework, education and awareness raising, financing schemes, a suitable IT infrastructure, a safe and profitable recovery of materials at the end-of-use stage, the support of circular business creation, circular product design and reverse supply chains aided by ICTs.

An assessment of the country's readiness for a CE transition was carried out through the evaluation of every aspect included in the proposed enabling framework. The Colombian case study shows that an economy highly dependent on the extraction of non-renewables such as oil and coal can potentially lock-in any sustainability strategy. This economic development model based on natural resources extraction is not compatible with the new development narrative proposed by the CE, which promotes natural capital regeneration and finding new ways of doing businesses. Thus, it is necessary to develop a joint strategy between the economic and environmental plans to foster the creation of an industrial sector framed within the CE principles able to generate wealth sustainably. This will imply a reduction in taxes associated to jobs generation and taxing the extraction of natural resources, non-renewables and environmental damage. This also implies avoiding harmful subsides which support the extraction of non-renewable resources.

Furthermore, the assessment of the country's digital infrastructure and use of ICTs in the industrial sector have indicated that more investment must be done to increase the coverage and quality of broadband and 4G services as well as to support SMEs to adopt and innovate through ICTs as a means of generating new business models within a circular framework.

Colombia also presents opportunities to safely and profitably circulate the materials from municipal solid waste streams. Although waste collection rates are good and the country has been implementing EPR schemes since 2011, recycling is mostly developed by the informal sector without the proper infrastructure or safe working conditions. It is also suggested to foster other inner-cycle activities such as repair, remanufacturing, cascading or reuse. These could create new business opportunities that avoid the disposing of waste into landfill.

An assessment of the innovation efforts, specifically those oriented to eco-innovation, shows that the Colombian large industrial enterprises could adopt design-led approaches to production, such as Cradle to Cradle, to innovate with sustainable products. Moreover, given the importance of SMEs in the creation of a new economic paradigm such as the CE, more technical and economic support should be provided to them. Some pilots to implement closed-loop approaches to production in this sector have shown that they lack the appropriate technical knowledge, the support of suppliers as well as financial mechanisms to implement changes oriented to a circular economy. In addition, it is important to showcase the economic and environmental benefits of a sustainability approach to incentivise companies to invest in these practices.

This chapter also suggests evaluating the effectiveness of recent initiatives to promote education and increase awareness about sustainability topics, as well as increase efforts to incorporate CE concepts and systems thinking in the university and school curricula as a basis to train human capital with a different mind-set.

References

Cornell University, INSEAD and WIPO (2016) *Global Innovation Index 2016: winning with global innovation. Ithaca, Fontainebleau, and Geneva*. [Online] Available at: http://www.wipo.int/edocs/pubdocs/en/wipo_pub_gii_2016.pdf

DNP (2014) *Bases del Plan Nacional de Desarrollo 2014–2018* [in Spanish]. [Online] Available at: http://colaboracion.dnp.gov.co/CDT/Prensa/Bases%20Plan%20Nacional%20de%20Desarrollo%202014-2018.pdf

Egerton-Read, S. (2016) Circular economy in emerging markets: an insight from Brazil. *Circulate: Emerging Markets*. [Online] Available at: https://circulatenews.org/2016/07/circular-economy-in-emerging-markets-an-insight-from-brazil-2

Ellen MacArthur Foundation (2013) *Towards the Circular Economy. Volume 1.* [Online] Available at: https://www.ellenmacarthurfoundation.org/assets/downloads/publications/Ellen-MacArthur-Foundation-Towards-the-Circular-Economy-vol.1.pdf

Ellen MacArthur Foundation (2015a) *Delivering the Circular Economy: a toolkit for policymakers*. [Online] Available at: https://www.ellenmacarthurfoundation.org/assets/downloads/publications/EllenMacArthurFoundation_PolicymakerToolkit.pdf

Ellen MacArthur Foundation (2015b) *Growth Within: a circular economy vision for a competitive Europe*. [Online] Available at: https://www.ellenmacarthurfoundation.org/assets/downloads/publications/EllenMacArthurFoundation_Growth-Within_July15.pdf

Ellen MacArthur Foundation (2015c) *Towards a Circular Economy: business rationale for an accelerated transition*. [Online] Available at: https://www.ellenmacarthurfoundation.org/assets/downloads/TCE_Ellen-MacArthur-Foundation_9-Dec-2015.pdf

European Commission (2014) Scoping study to identify potential circular economy actions, priority sectors, material flows and value chains. [Online] Available at: https://www.eesc.europa.eu/resources/docs/scoping-study.pdf

European Commission (2018) *2018 Circular Economy Package*. [Online] Available at: http://ec.europa.eu/environment/circular-economy/index_en.htm

Garcia, H. and Calderon, L. (2013) *The Political Economy of Fuel Subsides in Colombia*. OECD Environment Working Papers, No. 61. Paris: OECD Publishing. [Online] Available at: https://doi.org/10.1787/5k3twr8v5428-en

Garcia Caicedo, C.L. (2016) *Circular Economy Transition in the Context of Low and Middle-Income Countries: assessment of the circular economy transition readiness in Colombia*. Master's dissertation. University of Bath.

Geels, F.W., Elzen, B. and Green, K. (2004) General introduction: system innovation and transitions to sustainability. In Elzen, B., Geels, F.W. and Green, K. (eds.) *System Innovation and the Transition to Sustainability*, pp. 1–16. Cheltenham, UK and Northampton, MA: Edward Elgar Publishing Limited.

Ghisellini, P., Cialani, C. and Ulgiati, S. (2016) A review on circular economy: the expected transition to a balanced interplay of environmental and economic systems. *Journal of Cleaner Production*, 114, 11–32. [Online] Available at: http://dx.doi.org/10.1016/j.jclepro.2015.09.007

Global Footprint Network (2016) National footprint accounts: 2016. [Online] Available at: https://www.footprintnetwork.org/2016/03/08/national-footprint-accounts-2016-carbon-makes-60-worlds-ecological-footprint/

Gower, R. and Schöder, P. (2016) *Virtuous Circle: how the circular economy can create jobs and save lives in low and middle-income countries*. Teddington: Institute for Development Studies. [Online] Available at: https://www.ids.ac.uk/files/dmfile/TearfundVirtuousCircle-2.pdf

Haas, W., Krausmann, F., Wiedenhofer, D. and Heinz, M. (2015) How circular is the global economy? An assessment of material flows, waste production, and recycling in the European Union and the world in 2005. *Journal of Industrial Ecology*, 19 (5), 765–777.

Hobson, K. and Lynch, N. (2016) Diversifying and de-growing the circular economy: radical social transformation in a resource-scarce world. *Futures*, 82, 15–25.

Hoornweg, D. and Bhada-Tata, P. (2012) *What a Waste: a global review of solid waste management*. Urban Development Series, Knowledge Papers No. 15. Washington, DC: World Bank.

Lacy, P. and Rutqvist, J. (2015) *Waste to Wealth: the circular economy advantage*. 1. Basingstoke: Palgrave Macmillan.

Lieder, M. and Rashid, A. (2016) Towards circular economy implementation: a comprehensive review in context of manufacturing industry. *Journal of Cleaner Production*, 115, 36–51.

MADS (2010) *Politica Nacional de Producción y Consumo* [in Spanish]. Bogota: MADS. [Online] Available at: http://www.minambiente.gov.co/index.php/component/content/article/154-#pol%C3%ADticas

MarketLine (2015) *Country Profile Series: Colombia in-depth PESTLE insights*. London: Progressive Digital Media Ltd.

Medina, M. (2000) Scavenger cooperatives in Asia and Latin America. *Resources, Conservation and Recycling*, 31, 51–69.

Murray, A., Skene, K. and Haynes, K. (2015) The circular economy: an interdisciplinary exploration of the concept and application in a global context. *Journal of Business Ethics*, 140 (3), 369–380.

OCED/ECLAC (2014) *OECD Environmental Performance Reviews: Colombia 2014*. OECD Environmental Performance Reviews. Paris: OECD Publishing. [Online] Available at: https://doi.org/10.1787/9789264208292-en.

OECD (2014) Assessment and recommendations: towards a more efficient telecommunication sector in Colombia. In *OECD Reviews of Innovation Policy: Colombia 2014*. Paris: OECD Publishing, pp. 1–11. [Online] Available at: http://dx.doi.org/10.1787/9789264208131-en.

OECD (2015) *OECD Economic Surveys: Colombia 2015*. Paris: OECD Publishing. [Online] Available at: http://dx.doi.org/10.1787/eco_surveys-col-2015-en

OECD (2016) *Extended Producer Responsibility: updated guidance for efficient waste management*. Paris: OECD Publishing. [Online] Available at: https://doi.org/10.1787/9789264256385-en

Preston, F. (2012) *A Global Redesign? Shaping the circular economy*. [Online] Available at: http://www.chathamhouse.org/sites/files/chathamhouse/public/Research/Energy,Environment and Development/bp0312_preston.pdf

Skene, K. and Murray, A. (2015) *Sustainable Economics: context, challenges and opportunities for the 21st-century practitioner*. Sheffield: Greenleaf Pub. Ltd.

Stahel, W.R. (2016) Circular economy. *Nature*, 531 (24 March 2016), 435–438.

Stahel, W.R. and Clift, R. (2016) Stocks and flows in the performance economy. In Clift, R. and Druckman, A. (eds.) *Taking Stock of Industrial Ecology*, 137–158. Springer International Publishing.

Su, B., Heshmatia, A., Geng, Y. and Yu, X. (2013) A review of the circular economy in China: moving from rhetoric to implementation. *Journal of Cleaner Production*, 42, 215–227.

Tukker, A. (2015) Product services for a resource-efficient and circular economy: a review. *Journal of Cleaner Production*, 97, 76–91.

UNEP (2006) *Circular Economy: an alternative model for economic development*. [Online] Available at: http://www.unep.fr/shared/publications/pdf/DTIx0919xPA-circulareconomyEN.pdf

United Nations Development Program (2015) *Human Development Report 2015*. [Online] Available at: http://hdr.undp.org/sites/all/themes/hdr_theme/country-notes/MEX.pdf

United Nations Environment Program (2016) *GEO-6 Regional Assessment for Latin America and the Caribbean, Nairobi, Kenya*. [Online] Available at: http://www.unep.org/publications/

Vallejo, M.C., Pérez Rincón, M.A. and Martinez-Alier, J. (2011) Metabolic profile of the Colombian economy from 1970 to 2007. *Journal of Industrial Ecology*, 15 (2), 245–267.

von Weizsäcker, E.U., de Larderel, J., Hargroves, K., Hudson, C., Smith, M. and Rodrigues, M. (2014) *Decoupling 2: technologies, opportunities and policy options*. [Online] Available at: www.resourcepanel.org/sites/default/files/documents/document/media/-decoupling_2_technologies_opportunities_and_policy_options-2014irp_decoupling_2_report-1.pdf

Wilson, D., Velis, C. and Cheeseman, C. (2006) Role of informal sector recycling in waste management in developing countries. *Habitat International*, 30 (4), 797–808.

WWF (2015) EU Circular economy package: a failed promise. [Online] Available at: www.wwf.eu/?257498/EU-circular-economy-package-a-failed-promise

Bigano, A., Sniegocki, A. and Zotti, J. (2016) Policies for a more dematerialized EU economy: theoretical underpinnings, political context and expected feasibility. *Sustainability*, 8 (8), 717.

8
PROMOTING INDUSTRIAL SYMBIOSIS IN CHINA'S INDUSTRIAL PARKS AS A CIRCULAR ECONOMY STRATEGY

The experience of the TEDA Eco Centre

An Chen, Yuyan Song and Kartika Anggraeni

Introduction

The circular economy (CE) is an economy that aims at retaining products, components and materials at their highest level of utility and value (Saavedra et al., 2018). It is about a continuous positive cycle of development in which natural capital is 'conserved and enhanced, minimising systemic risk by managing finite stocks and renewable flow' (Saavedra et al., 2018). However, some critics have said that CE means many different things to different people (Kirchherr, Reike and Hekkert, 2017). In China, CE has been perceived as an embodiment of ecological civilisation (*sheng tai wen ming*), which attempts to overcome conflicts between the environment and economic growth through innovations, both technical and social. Since China's rapid economic growth has created severe environmental impacts, the CE is framed as a new model to reconcile this problem (Liu and Côté, 2017; McDowall et al., 2017), whereas in the European context, the CE is framed as a way of turning environmental necessity into economic opportunity (McDowall et al., 2017). The specific CE approach of industrial symbiosis (IS) is often seen by industrial ecologists as 'part of a CE that focuses on keeping the added value in products for as long as possible and at the same time eliminating waste' (Nordregio Report, 2015).

Circular economy in China

The CE was formally accepted in 2002 by China's central government following the country's rapid economic growth, which resulted in a shortage of raw materials and energy (Yuan, Bi and Moriguichi, 2006) and lead to economic as well as ecological costs (Mathews and Tan, 2011; Liu and Côté, 2017). In 2005, China's State Council issued the CE Acceleration Policy (China's State Council, 2005), in which the National Development and Reform Commission (NDRC) jointly with five other ministries initiated the first batch of CE pilots (NDRC, 2016). The initiative included 42 companies from seven key industries, including steel, non-ferrous metals, coal, power, chemical, building materials and lighting. Furthermore, 17 recycling companies from four key fields (construction resource recycling system, scrap metal recycling, waste household appliance recycling, and manufacturing) were involved, as were 13 industrial parks (IPs)[1] and ten resource-based or resource-scarce cities major urban centres such as Beijing, Shanghai, Liaoning, Jiangsu, Shandong and Chongqing. The goal was to explore strategic CE models that can be replicated across the country. After this pilot initiative, in 2009 the Chinese government passed the Circular Economy Promotion Law, making CE an official development goal (Liu and Côté, 2017; Mathews and Tan, 2011; Yuan, Bi and Moriguichi, 2006).

Circular economy and China's industrial parks

IPs have become China's engines of economic growth over the last 30 years. Liu and Côté (2017) suggest considering IPs as CE's microcosms, and Shi and Chertow (2017) put forward that eco-industrial parks (EIPs) are a key vehicle for IS implementation. To date there are 2,543 IPs at the national and provincial levels (NDRC, 2018) and out of these, 365 are national-level IPs, which are called 'National Development Zones' (NDZs).[2] In 2016, these NDZs generated a turnover of £1.9 trillion accounting for 23 per cent of China's GDP (Tongji University, 2018). IPs also account for a large share of manufacturing output (McDowall et al., 2017), thus have a huge potential in contributing to implementing CE elements.

IPs offer companies cost efficiency through a joint use of infrastructures (e.g. roadways, railroad sidings, ports, high-power electric supplies, heating and water supplies). However, with the concentrated industrial activities, IPs have become hotspots of intensive consumption of resources and energy and the main sources of pollutant discharges due to inefficient resource use and wasteful production models. China still requires 2.5 kilogrammes of materials to generate USD 1 of GDP compared to 0.54 kilogrammes in OECD countries (Mathews & Tan, 2016). There are still a number of IPs, especially in less developed areas like central and western regions of China, which are facing great challenges in energy and environmental management. The IP administrative governments and enterprises still lack the following: i) awareness of CE as a strategy to achieve sustainable development, ii) knowledge about environmental risk control, iii) capacity in environmental and energy management, and iv) information about clean technologies and the importance of their adoption.

Industrial symbiosis to realise circular economy within industrial parks

Implementing IS within IPs has been considered as a strategy toward realising a CE (McDowall et al., 2017) and achieving sustainable growth. With IS, one understands 'physical exchanges of materials, energy, water, and by-products among diversified clusters of firms' (Chertow, 2007). Through symbiotic activities between firms, IS promotes sustainable resource use at the inter-firm level, thereby minimising the need for using virgin raw materials and at the same time eliminating waste (Nordregio Report, 2015). However, there still is a lack of understanding of how industrial symbiosis can be developed and encouraged by public institutions and private enterprises.

This chapter draws on the experience of the TEDA Eco Centre (TEC) in promoting CE. It facilitated industrial symbiosis between companies and partnerships with local government and TEDA authorities. Since TEDA has been featured as a case study in various literatures as successful in promoting environmental protection in industrial parks (UNIDO, 2016; Shi and Yu, 2014; Mathews & Tan, 2011), this chapter presents a timely opportunity to learn from the TEC's own perspective and experience in facilitating industrial symbiosis since 2010.

TEDA Eco Centre

The TEDA Eco Centre (TEC) is a non-profit state-sponsored organisation which was founded in 2010 through a cooperation between the Administrative Committee of Tianjin Economic-Technological Development Area (TEDA) and the European Union (EU). Initially, the role of TEC was to promote CE and low-carbon transformation within TEDA. To this end, TEC served as an information sharing platform, solving the issue of information asymmetry between companies. It collected and shared information on the types and amounts of wastes produced within TEDA and promoted the uptake of relevant clean technology. TEC organised business matchmaking, SME trade shows and outreach as well as facilitated cooperation with international organisations to promote knowledge exchange. TEC offered companies consultation and training such as environmental and occupational health and safety (OHS) management systems (ISO14001 and 45001), energy/carbon auditing, and environmental information disclosure. After the completion of the project, TEC continued operations and extended its services to other IPs, thus contributing to a national low-carbon and CE development.

TEDA and industrial symbiosis

The CE, as a counter-measure to resource bottlenecks and adverse environmental impacts, has attracted significant attention among IPs since 2005 (China's State Council, 2005). Following the issue of the CE Promotion Law in 2009, TEDA, as one of the earliest National Economic and Technological Development Zones, embarked on a journey of identifying CE models that work. In fact, the focus on transforming TEDA towards an eco-industrial park and applying CE elements at the industrial park level had started in the early 2000s (Zhu, 2009). For example, a cogeneration power station was built in 2003, using treated wastewater as boiler supply water (Mathews and Tan, 2011). TEDA had established several IS networks based on its pillar industries and main production resources, such as automobile industry (see Figure 8.1) and water resources (see Figure 8.2). In the industrial network of the automobile industry, key raw materials like aluminium waste, steel waste and lead waste are collected and recycled by recycling companies located within TEDA, and put back into manufacturing processes to produce castings.

With support from the EU-funded SWITCH-Asia Programme (2009–2013), the TEDA authority and Environmental Protection Agency (EPA) implemented a project with the title Industrial Symbiosis (IS). The project started developing the IS network, but with slow progress due to lack of expertise and manpower in identifying synergies among hundreds of companies. Therefore, TEDA established the Eco Centre in 2010 to facilitate symbiotic relationships between companies (Figure 8.3). At the same time, it cooperated with the UK's National Industrial Symbiosis Programme (NISP) to replicate NISP's proven and network-based CE service model to three industrial parks located in Tianjin Binhai New Area (TBNA).[3] TBNA is a 2,270 km^2 cluster of industrial zones with the highest concentration of industrial output in Northern China. In the IS project, TEDA worked with its two neighbouring IPs, which generated huge amounts of solid waste and environmental pollution. At that time, companies located in these three IPs had weak environmental management capacity with no networks available to facilitate synergies among the companies located within the IPs.

With the initial support from international actors such as NISP and UNIDO, TEC collected information of companies' waste and by-product streams and facilitated an exchange of secondary raw materials, such as solid waste, excess energy, and other by-products. TEC also engaged with local governments, i.e. the authorities of the three IPs, which helped to create buy-in from the companies located within the IPs and encourage them to share data on their waste and by-products. All this was done by TEC while maintaining its neutral stance as a third party. Through its work, TEC proved that the involvement of local governments was instrumental in accelerating the development of an IS network. Local governments lent the IS project credibility and helped advertise services offered by TEC to companies. By setting up common goals and establishing TEC as a service provider to industries, the local governments attracted a large number of companies to join the IS network within TBNA.

The IS project and its results

The IS project offered a one-day environmental management audit and training on waste and by-product data collection methods. During 2010–2013, TEC organised eight environmental management trainings attended by 433 participants from 300 companies (see Table 8.1). Data was collected in a standardised format and entered into a comprehensive database which included companies' material inputs and outputs, types of materials and wastes, quantity, processes, production needs and capacities, as well as utilities and logistics. Between 2010 and 2012, 574 investigation sheets from various companies were collected. As facilitator, TEC ensured the confidentiality of the shared data through signing agreements with concerned

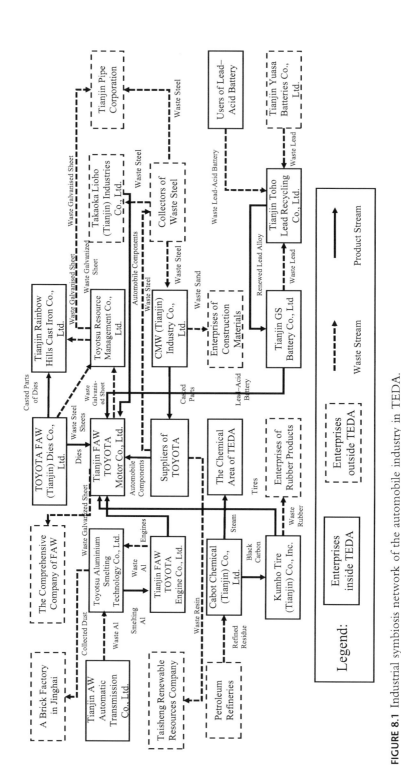

FIGURE 8.1 Industrial symbiosis network of the automobile industry in TEDA.
Source: TEDA Eco Centre, 2015.

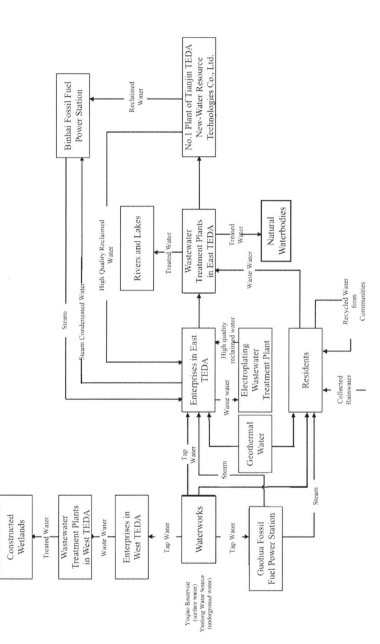

FIGURE 8.2 Industrial symbiosis network for water resources in TEDA.
Source: TEDA Eco Centre, 2015.

companies. This database has enabled the matching of supply and demand of secondary materials within the IS networks. As a result, potential synergies were identified and confirmed through direct communication between interested parties in workshops or through expert analysis and recommendations. Figure 8.4 presents an overview of this workflow.

An example of a synergy was the case of Samsung Tianjin and Taiding. Samsung is a producer of digital cameras, and its factories in TEDA are primarily manufacturing digital cameras, video recorders and other camera devices which are sold to the entire Asian market. However, at the time of the IS project implementation, it encountered a problem with waste plastic housing, to which a recycling company, Taiding, offered a solution. With TEC's facilitation, Samsung and Taiding established a long-term cooperation. Taiding recycled 8 tonnes of plastic housing for Samsung and through this reduced about 103 tonnes CO_2 emission and saved 8 tonnes of materials which would otherwise have gone to landfill.

The IS project concluded in 2013 with 99 new synergies (one synergy involved at least two companies) (see Figure 8.5). The synergies diverted 1.43 million tonnes of solid waste from landfills, reduced 167,000 tonnes of CO_2 emissions and saved production costs of £8.2 million. Over 100 SMEs participated in quick environmental management audits, and 41 SMEs followed it through and adopted a full environmental management system (EMS), and eventually obtained an ISO14001 certification.

Source: TEDA Eco Centre, 2014.

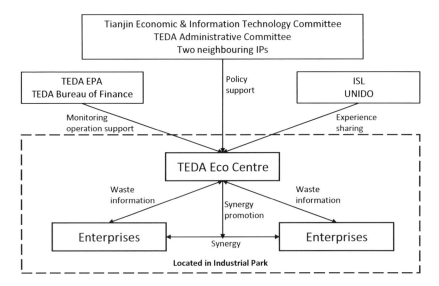

FIGURE 8.3 TEDA Eco Centre as waste information exchange/sharing platform in the Industrial Symbiosis project.

Source: TEDA Eco Centre, 2018.

At the time of writing, five years after the project, only a handful of the identified symbioses still continue. This is partly due to the companies' preference to simply sell their waste (metal scraps, plastics, paper etc.) to collecting/recycling companies, rather than finding other companies to exchange secondary materials or by-products. This approach to waste is deemed more profitable and practical by manufacturing companies since it does not require operational changes and takes less human resources and time.

Another reason is quite fundamental: the flourishing industry of non-hazardous waste collection and waste recycling in China. Nowadays, small companies can invest little money to establish 'waste collection stations' and earn some margins resulting from the difference between prices set by the manufacturing plants and recycling companies. This has created a solid business model for non-hazardous waste, i.e. the manufacturing plants sell secondary materials to the collection stations, which in turn sell these to the recycling companies. In short, sending non-hazardous solid waste to collection and recycling facilities is easier than finding other companies that want to use the secondary materials for their production. The rest of solid waste which is categorised as hazardous is generally sent to a professional hazardous waste treatment plant for incineration or recovery.

Labelling and TEDA's new model of waste management

TEC also implemented the Whole Process Management of General Industrial Solid Waste project (Figure 8.6). The project's objective was to create a general solid waste manifest system[4] by applying a methodology adopted from Japan to track waste streams from production to storage, transportation, recycling, reuse and

TABLE 8.1 ISO 14001 Environment Management System training during the IS project.

No.	Date	Regions	Training site	No. of enterprises	No. of attendees
1	June 22, 2010	TEDA, THIP	TEC	82	121
2	June 24, 2010	TEDA, THIP	TEC		
3	July 5, 2010	TPFTZ	TPFTZ	44	68
4	May 19, 2011	TEDA	TEC	45	71
5	July 21, 2011	Motorola suppliers	Motorola plant	43	43
6	March 31, 2012	TEDA	TEC	44	66
7	July 26, 2012	TEDA	TEC	35	42
8	May 29, 2013	TBNA	TEC	7	22

Source: TEDA Eco Centre, 2014.

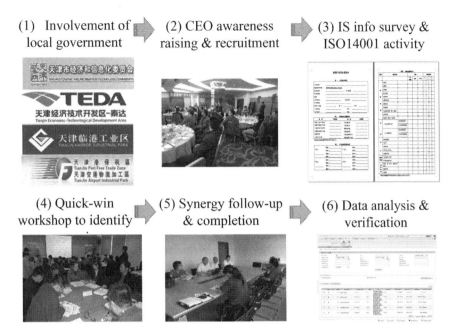

FIGURE 8.4 Industrial symbiosis workflow in the TBNA case.
Source: TEDA Eco Centre, 2018.

final treatment. In order to map the waste streams, the project promoted a TEDA eco-management label specifically to three categories of companies, i.e. manufacturing plants within TEDA, collecting companies, and recycling companies within or outside TEDA. In 2014, 32 manufacturing plants and 16 collecting/recycling companies had obtained the label.

During the project, TEC conducted an on-site audit to evaluate the companies' performance on waste management. When a company could obtain the eco-management label for three consecutive years, it would be awarded with a subsidy of around £3,400 by the TEDA authority. The subsidy was provided under the TEDA Special Fund of Energy Saving-Consumption Reducing and Environment Protection (TEDA, 2016). This special fund has an annual budget of around £11.3 million and was established in 2007 by the TEDA authority as a long-term policy to encourage local companies adopting energy-saving and environmental protection strategies. Under this fund, TEDA authority regularly (every 1–2 years) updates and issues a subsidy inventory that lists the categories of energy-saving and environmental protection measures, which conform to the new requirements of governmental policies as well as developing sustainability trends. The labelling system has helped the manufacturing plants to identify and contract collection and recycling companies that performed well on waste management. This has considerably improved the level of non-hazardous solid waste management and laid the groundwork for a transparent recycling market in TEDA.

Synergies resources	Number of synergies
Plastic	33
Paper	23
Metal	21
Organics	9
Wood	3
Sludge	4
Limestone & gypsum	5
Reclaimed water	1
Total	99

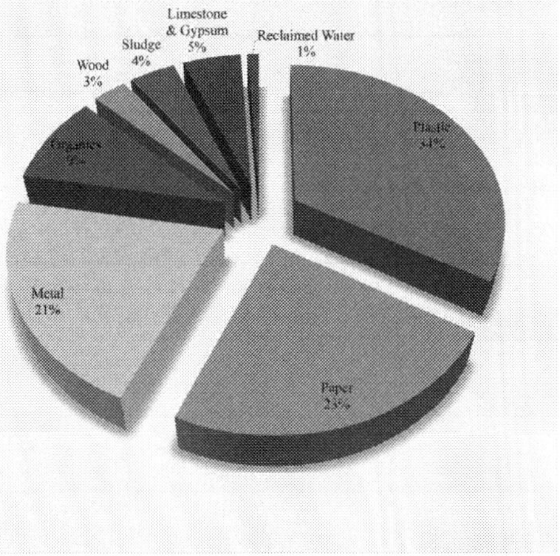

FIGURE 8.5 Types of waste material and number of related synergies created through the IS project.

In 2012, the project managed to track 0.175 million tonnes of general non-hazardous solid waste from 25 pilot manufacturing plants and collected 1,456 manifests containing data on waste generated by these plants. Collecting and recycling companies were also involved in the tracking activities. Using the information, TEC mapped the waste streams and was able to track various waste streams, including transfer, treatment and recycling of waste within as well as outside TEDA (Figure 8.7).

During the on-site audit of companies, TEC auditors used a set of indicators to assess the companies' performance in waste management. If it turned out that a company was non-compliant, the auditors provided consultancy to the company on how to improve its waste management practice. Annually, in April or June TEC releases an updated list of collecting and recycling companies having the eco-management label to the manufacturing plants within TEDA. The manufacturing plants can find a list of certified collection and recycling companies, in case they need to find new partners.

In 2016, the Whole Process Management of General Industrial Solid Waste project and the IS project were included as one of nine exemplary showcases in the Notice of NDRC (2016) and the Ministry of Finance. The Notice recognised the nine projects as national circular economy pilots and demonstrations. Hence, the manifest and labelling systems as well as industrial symbiosis have been introduced to other IPs in China, such as Qinhuangdao Economic and Technological Development Zone.

Overcoming obstacles towards CE and public awareness raising

Several obstacles hindering the promotion of CE development within companies were identified by TEC, including a low level of awareness, a lack of budget for CE promotion and training, and limited access to clean technology or solutions to

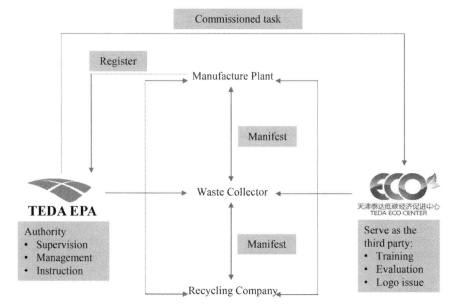

FIGURE 8.6 Whole Process Management of General Industrial Solid Waste project in TEDA.
Source: TEDA Eco Centre, 2015.

enhance energy and resource efficiency. To address the problems, TEC implements several measures as discussed in the following.

Enhancing public awareness and facilitating technology transfer

TEC facilitated the transfer of advanced clean technologies. For example, TEC received funding from China's National Development and Reform Commission (NDRC) to conduct a project, namely TEDA Promotion Base of Circular Economy Technology, in 2013. The project's objective was to bring technology suppliers of CE and low carbon solutions to exhibit and promote their products within TEDA and the interested public of Tianjin city. TEC keeps its database of CE-related technologies updated and organises 3–4 matchmaking events per year to promote investments, for example, in surplus energy reclamation, water recycling, and plant-wide energy saving (Figure 8.8). To create public awareness and build the capacity of workers and residents in TEDA, TEC conducts at least eight seminars and trainings on green development and CE every year.

Launching CE information service platforms

Earlier in 2013, TEC launched an industrial symbiosis online platform to help identify and realise synergies between companies in TEDA. However, the platform failed due to the limited number of users, a lack of good trading mechanism as well as a low frequency of waste trading. The latter was a result of the already existing contracts between manufacturing plants and collection and recycling companies. This discouraged companies to seek new partnerships. In 2016, a new CE information service platform was launched with funding from NDRC. This platform targets three functional entities: enterprises, industry chains, and industrial parks. Also, it provides tools to help companies and government in their decision-making. The new online platform benefits industries at three levels: i) on the company level, as the online platform provides a search function to identify potential IS synergies and find quick matches; ii) on the industrial chain level, because the platform offers an industrial model, a symbiotic algorithm, and a trade module to facilitate an inter-enterprise IS networks; iii) on the industrial park level, a visualisation module, a material flow calculation module, and a smart management module help to create comprehensive CE management in industrial parks. TEC plans to expand this platform to also offer circular technology modules which support companies in deciding which technology to invest in. In the same year, TEC reported that 248 companies had used the CE information platform to exchange information, which eventually improved their management of raw materials and wastes. By its definition, despite the existence of such an information exchange platform, industrial symbiosis has not yet taken place. It can happen only when companies start exchanging their 'wastes' such as by-products and wastewater by using information provided by the platform.

Replicating the TEC model for mainstream CE and industrial symbiosis in other industrial parks

Evidently, TEC has become the 'green node' of TEDA. It is gathering information about waste, clean-tech, external experts, policies, and finance, and shares this information with companies located within and around TEDA. TEC also develops its functions and offers more sophisticated consulting and training services, such as environment compliance evaluations and clean-tech market analysis. These additional services are necessary to identify the needs of companies, resulting in good channels of gathering information from the demand side. TEC has subsequently established a functional CE and low-carbon information exchange mechanism for the Chinese context (Figure 8.9), which can be replicated in other IPs seeking to embrace the circular economy.

Due to its achievements in promoting the green development of industrial parks, TEC has become the green development secretariat of the League of National Economic-Technological Development Zones (NETDZs). The League was established in 2016 and consists of 219 national-level industrial parks in China, under the Ministry of Commerce. Since then TEC has served hundreds of IPs nationwide and launched the Green Development Information Platform for NETDZs to release and share information. This includes information about IPs, policy and regulations relevant to IP green development, 'eco-friendly' companies providing services to IPs,

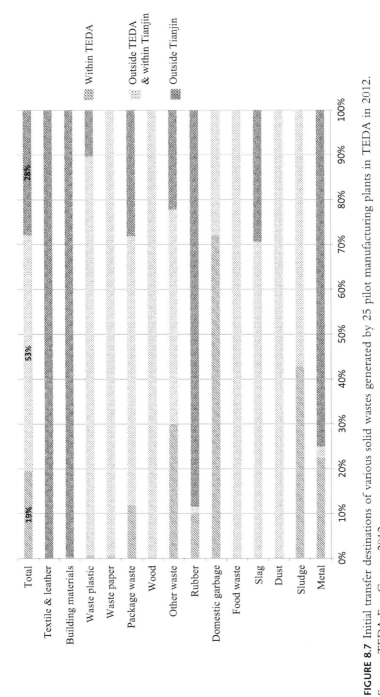

FIGURE 8.7 Initial transfer destinations of various solid wastes generated by 25 pilot manufacturing plants in TEDA in 2012.
Source: TEDA Eco Centre, 2012.

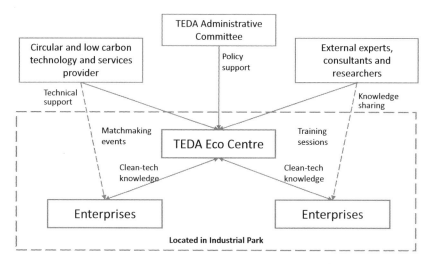

FIGURE 8.8 TEDA Eco Centre as clean-tech and knowledge promotion platform.
Source: TEDA Eco Centre, 2018.

and experts in IP planning and management. The platform also covers other topics, such as low carbon IPs, EIPs, and smart IPs. Further, TEC has established an Expert Committee. Since the foundation of the League, TEC has organised various trainings on IP green and circular development, attracting 1,200 governmental officers from 160 national IPs and 100 provincial IPs.

Discussion: how to move the CE beyond recycling?

The case of TEC shows the importance of establishing an eco centre in industrial parks to facilitate partnerships between industries and local authorities, as well as among companies, in realising a CE and green industrial development. Some of the literature suggests that in order to build trusts (Ruiz Puente, Romero Arozamena and Evanset, 2015) and social connections among companies first within a particular IP and later among IPs, a 'coordination team' (Trevisan et al., 2016), a supporting organisation (Liu and Côté, 2017) or a third-party facilitator is needed (Velenturf, 2017), as was showcased by TEC playing the role of a facilitator for TEDA. Two projects implemented by TEC have been made 'national showcases' by the Chinese government, where the manifest and labelling systems were replicated in other IPs. This success is due to TEC's role in integrating the local government's policies (on energy efficiency and environmental protection) into its service packages to companies located in TEDA. By taking TEC's service packages, companies can ensure compliance with the regulations, and to a certain degree, going beyond the compliance.

The Industrial Symbiosis (IS) project and the Whole Process Management of General Industrial Solid Waste project have showcased the huge potential of IS in enhancing resource efficiency, reducing waste and pollutions, as well as achieving a low carbon economy. The emphasis of achieving CE through efficiency (Zhu, 2009) has always been TEDA's focus. TEDA has a list of key projects for energy saving and the development of recycling economy. The key projects reflected TEDA's regional environmental protection and industrial development policies (Zhu, 2009). This may explain why saving energy, constructing clean energy systems, and the use of clean and renewable energy in TEDA seem to take precedence over IS.

Despite the Chinese government's CE policy and the related waste trading as described in its 13th Five-Year Plan (2016–2020), the emphasis was more on *reducing pollution* and a broader definition of sustainable development and ecological civilisation rather than on waste and opportunities for industry (McDowall et al., 2017). Thus, in the case of TEDA, most companies seek to reduce their polluting operations instead of attempting to build synergies with other companies, which would not only reduce their waste and improve the environment, but would also require a certain degree of innovation and collaboration between companies for long-term social and economic benefits.

The recycling industry also seemed to pose a barrier to IS. Collection and recycling companies have recently been mushrooming in China and aggressively obtain secondary materials from any possible sources within a certain distance. This trend encourages companies to sell their still-valuable waste to recycling companies and send the rest to landfills. Thus, the IS implementation by companies within TEDA can still be further improved. Companies might still have limited awareness of the environmental benefits of industrial symbiosis and motivation (IISD, 2015).

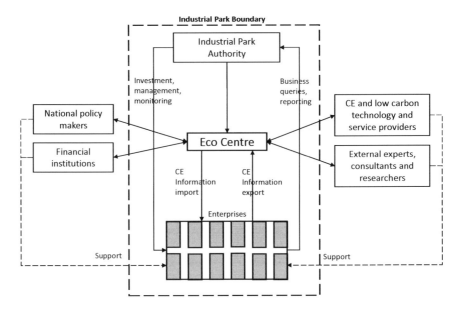

FIGURE 8.9 Eco centres as circular economy and low-carbon information exchange/sharing platforms to promote CE development in industrial parks.
Source: TEDA Eco Centre, 2018.

Further, the limited replication of IS within TEDA or to other IPs might be caused by a lack of CE clear guidance and definition on the part of IP administrations. Lack of capacity building and limited technological development still hamper eco-efficient industrial parks in China (IISD, 2015). Hence, only with supporting policies from local, provincial and national governments that provide clear guidance and incentives to companies can IS be widely promoted (Liu and Côté, 2017).

It is clear that government plays a huge role as an enabler at the national level and the IP level (Shi and Yu, 2014). It has an indispensable role in 'governance, policy-making and providing support systems for the development of business' (Liu and Côté, 2017). Therefore, to ensure environmental protection and CE integration within IPs in China, the experience of TEDA confirms that it is necessary to build 'a mechanism for executing a fee system for resources and eco-compensation, reform the prices of natural resources and products in order to fully reflect market supply and demand, resource scarcity, environmental damage cost and ecological restoration' (Liu and Côté, 2017).

Conclusions

Industrial parks can be regarded as microcosms of the circular economy. They have a vital role to play to support China's aim of green industrial development and more sustainable growth. The example of the TEDA Eco Centre (TEC) shows the importance of establishing an eco-centre within every industrial park as an intermediary organisation to organise and facilitate partnerships and industrial symbiosis networks between companies and local governments. In the context of international development cooperation programmes, TEC has been successful in establishing mechanisms and platforms to promote industrial symbiosis, which are key to the circular economy. However, relatively few industrial synergies have continued beyond the projects, partly due to the booming recycling industry and lack of clear guidance and incentives for companies to pursue industrial symbiosis. Transporting waste requires fuels (energy) and produces extra CO_2 emissions, besides the fact that many of the collection and recycling companies still use conventional technologies which further contribute to the environmental impacts. The question of whether waste recycling is more efficient than industrial symbiosis, in terms of environmental and economic impacts, needs to be further researched.

Notes

1 The industrial parks includes various development zones, e.g. Economic Development Zones, Hi-tech Industrial Development Zones, National Economic and Technological Zones, Bonded Areas, Export Processing Zones, Border Economic Cooperation Zones.
2 365 national development zones include 146 National Hi-tech Industrial Development Zones and 219 National Economic and Technological Development Zones.

3 The three IPs are the Tianjin Economic-Technological Development Area (TEDA), Tianjin Harbor Industrial Park (THIP), and Tianjin Port Free Trade Zone (TPFTZ).
4 A manifest systems in waste management is a pollution control procedure in which a hazardous material is identified and traced from its production, treatment, transportation, and final disposal through a series of linked documents called manifests.

References

China's State Council (2005) *Several Opinions of the State Council on Accelerating the Development of Circular Economy* [in Chinese]. [Online] Available at http://www.gov.cn/gongbao/content/2005/content_64318.htm [Accessed 12 July 2018].

Chertow, M.R. (2007) 'Uncovering' industrial symbiosis. *Journal of Industrial Ecology*, 11 (1), 11–30. [Online] Available at: https://doi.org/10.1162/jiec.2007.1110

Ellen MacArthur Foundation (2016) Intelligent assets: unlocking the circular economy potential. [Online] Available at: https://www.ellenmacarthurfoundation.org/assets/downloads/publications/EllenMacArthurFoundation_Intelligent_Assets_080216.pdf [Accessed 19 August 2018].

IISD (2015) *China's Low-Carbon Competitiveness and National Technical and Economic Zones: development of eco-efficient industrial parks in China (a review)*. Manitoba: IISD.

Kirchherr, J., Reike, D. and Hekkert, M. (2017) Conceptualizing the circular economy: an analysis of 114 definitions. *Resources, Conservation and Recycling*, 127 (2017), 221–232.

Liu, C. and Côté, R. (2017) A framework for integrating ecosystem services to China's circular economy: the case of eco-industrial parks. *Sustainability*, 9 (1510). [Online] Available at: https://doi.org/10.3390/su9091510

Mathews, J.A. and Tan, H. (2011) Progress toward a circular economy in China: the drivers (and inhibitors) of eco-industrial initiative. *Journal of Industrial Ecology* [e-journal], 15 (3), 435–457. [Online] Available at: https://doi.org/10.1111/j.1530-9290.2011.00332.x

Mathews, J.A. and Tan, H. (2016) Circular economy: lessons from China. *Nature*, 531 (7595), 440–442. [Online] Available at: https://doi.org/10.1038/531440a

McDowall, W., Geng, Y., Huang, B., Barteková, E., Bleischwitz, R., Türkeli, S., Kemp, R. and Doménech, T. (2017) Circular economy policies in China and Europe. *Journal of Industrial Ecology*, 21 (3), 651–661. [Online] Available at: https://doi.org/10.1111/jiec.12597

NDRC (2016) *Notice of the Typical Experience of Pilot Demonstration of National Circular Economy*. [Online] Available at: http://www.ndrc.gov.cn/zcfb/zcfbtz/201605/t20160510_801123.html [Accessed 15 June 2018].

NDRC (2018) *China Development Zone Audit Announcement Catalogue (Notice #4)*. [Online] Available at: http://www.ndrc.gov.cn/gzdt/201803/t20180302_878800.html [Accessed 15 June 2018].

Nordregio Report (2015) *The Potential of Industrial Symbiosis as a Key Driver of Green Growth in Nordic Regions*. Stockholm: Nordic Centre for Spatial Development.

Ruiz Puente, M.C., Romero Arozamena, E. and Evans, S. (2015) Industrial symbiosis opportunities for small and medium sized enterprises: preliminary study in the Besaya region (Cantabria, Northern Spain). *Elsevier Journal of Cleaner Production*, 87 (2015), 357–374. [Online] Available at: http://dx.doi.org/10.1016/j.jclepro.2014.10.046

Saavedra, Y.M.B., Iritani, D.R., Pavan, A.L.R. and Ometto, A.R. (2018) Theoretical contribution of industrial ecology to circular economy (review). *Journal of Cleaner Production*, 170 (2018), 1514–1522. [Online] Available at: https://doi.org/10.1016/j.jclepro.2017.09.260

Shi, L. and Chertow, M. (2017) Organizational boundary change in industrial symbiosis: revisiting the Guitang Group in China. *Sustainability*, 9 (1085). Available at: https://doi.org/10.3390/su9071085

Shi, L. and Yu, B. (2014) Eco-industrial parks from strategic niches to development mainstream: the cases of China. *Sustainability*, 6, 6325–6331. [Online] Available at: https://doi.org/10.3390/su6096325

TEDA (2016) *Interim Provisions on Promoting Energy Conservation, Consumption Reduction and Environmental Protection in Tianjin Economic and Technological Development Zone* [in Chinese]. [Online] Available at: http://www.teda.gov.cn/contents/690/1977.html [Accessed 27 July 2018].

Tongji University (2018) *Exploration of Legal System, System and Mechanism in New Era Park*. [Online] Available at http://tdi.tongji.edu.cn/News_show.asp?id=895 [Accessed 15 June 2018].

Trevisan, M., Nascimento, L.F., da Rosa Gama Madruga, L.R., Muelling Neutzling, D., Figueiro, P.S. and Bossle, M.B. (2016) Industrial ecology, industrial symbiosis and industrial eco-park: to know to apply. *Systems & Management*, 11 (2), 204–215. [Online] Available at: https://doi.org/10.20985/1980-5160.2016.v11n2.993

UNIDO (2016) *Global Assessment of Eco-Industrial Parks in Developing and Emerging Countries*. [Online] Available at: https://www.unido.org/sites/default/files/2017-02/2016_Unido_Global_Assessment_of_Eco-Industrial_Parks_in_Developing_Countries-Global_RECP_programme_0.pdf [Accessed 15 June 2018].

Velenturf, A.P.M. (2017) Initiating resource partnerships for industrial symbiosis. *Regional Studies Regional Science* [e-journal], 4 (1), 117–124. [Online] Available at: https://doi.org/10.1080/21681376.2017.1328285

Yuan, Z., Bi, J. and Moriguichi, Y., (2006) The circular economy: a new development strategy in China. *Journal of Industrial Ecology*, 10 (1–2), 4–8.

Zhu, T. (2009) Circular economy promoting the development of industrial parks in China: case study of TEDA. In: Holländer, R., Wu, C. and Duan, N. (eds.) *Sustainable Development of Industrial Parks (Working Paper No. 81)*, pp. 5–8. Leipzig: University of Leipzig.

9

ACCELERATING THE TRANSITION TO A CIRCULAR ECONOMY IN AFRICA

Case studies from Kenya and South Africa

Peter Desmond and Milcah Asamba

Introduction

The circular economy (CE) involves the redesign of products so that they are repairable and longer-lasting and re-used, failed components are replaced, core elements are refurbished, and precious metals and rare earths are extracted and remanufactured into new products. This approach can keep waste away from landfill while recovering high-value resources. There are many definitions of CE; 114 were studied by Kirchherr, Reike and Hekkert (2017). The authors proposed a definition, bringing together the many strands of their research:

> CE is an economic system that replaces the "end-of-life" concept with reducing, alternatively reusing, recycling and recovering materials in production/distribution and consumption processes. It operates . . . with the aim of accomplishing sustainable development, thus simultaneously creating environmental quality, economic prosperity and social equity, to the benefit of current and future generations. It is enabled by novel business models and responsible consumers.
>
> *(Kirchherr, Reike and Hekkert, 2017:229)*

In the Global North, particularly Europe, there is a greater understanding of circular practices in multinational organisations with well-documented case studies. Examples include Philips ('Pay Per Lux'), Desso and Interface (renting carpet tiles) and Renault (remanufacturing) (Ellen MacArthur Foundation, 2013a). The narrative in the EU and Global North is generally focussed around waste management through reuse and recycling and cost savings; the redesign of products and systems as well as new business models such as remanufacturing have only recently been emerging.

In the Global South, particularly India, China and South America, there are many small-scale examples of circular practices, e.g. waste collection and recycling, repair, refurbishment and biomass as a fertiliser in agriculture (Gower and Schröder, 2016). In Africa there is a greater emphasis on job creation, income generation and environmental impacts and on maximising the use of resources. For decades, circular activities have provided new and different kinds of skills and jobs, e.g. the vehicle repair and remanufacturing cluster in Kumasi in the Ashanti district of Ghana (Schmitz, 2015). In contrast, African case studies stay 'hidden' as they have yet to be documented through academic research. There is also a gap in research into how multinational businesses with global value chains (GVC) are engaged with small and medium-sized enterprises (SMEs) and entrepreneurs in Africa.

Several African countries are currently focusing on developing green economies which will feed into CE. UNEP defines the Green Economy (GE) as one that 'results in improved human well-being and social equity, while significantly reducing environmental risks and ecological scarcities'. In short, it is an economy that is low-carbon, resource efficient and socially inclusive. However, the concept and its implications are still vague when it comes to the African context (Klein et al., 2013). Different countries in Africa are at different stages of implementing GE, with some countries only integrating aspects of it while others, like Ethiopia, Kenya and Rwanda, have put in place an overarching GE strategy. The legal and regulatory framework to foster GE is still in its infancy stage in most African countries and mechanisms to realise the transition to GE are not yet in place. The most promising markets tend to be those related to agriculture, such as bio-trade, sustainable tourism and renewable energies (Klein et al., 2013).

By moving towards a CE strategy, emerging economies may be able to leapfrog to a more sustainable development approach by learning lessons and avoiding pitfalls of resource-intensive practices of the linear economy (Preston and Lehne, 2017). This chapter considers CE in an African context through policies and practices with a specific focus on Kenya and South Africa.

The circular economy in Africa

The current and projected increase of resource consumption in a globalised linear economy exceeds planetary boundaries (IRP, 2017). In addition, the redistribution of wealth from North to South continues to be essential for the 300 million people who live in poverty in countries still classified as low-income, mainly in sub-Saharan Africa (Raworth, 2017). CE has the potential to produce cost savings and reduce exposure to market price fluctuations, increasing renewable energy and releasing valuable materials and energy in existing products (Ellen MacArthur Foundation, 2013b). Yet this can be viewed by some actors as a risk and a threat. For example, governments which have become dependent on the export of natural resources are often reluctant to distribute the revenue generated more equally. Civil society has raised concerns about the health and environmental impacts of secondary materials such as e-waste that are imported into the country.

CE strategies in the North risk concentrating power and wealth amongst a few actors in global supply chains to the detriment of poor nations. For example, the European Commission's Circular Economy Action Plan (European Commission, 2017) identifies setting eco-design standards for electronic and electrical equipment, addressing hazardous chemicals in material cycles, and improving the circularity of plastics as priorities for Europe's transition to CE. Much of this plan focuses upon the benefits to Europe through greater resource efficiency. However, a more circular economy in Europe can also deliver benefits for people in low-income countries if their needs are better considered when creating inclusive CE policies. The more complex elements of restorative design and systemic change in CE are discussed (Webster, 2015) but generally only put into practice by businesses who can afford the cost impact of such changes.

In the past, many GVCs have relied upon Africa to provide virgin resources for the manufacture of products in the North (e.g. rare earths and minerals from DRC for the production of smartphones in China). CE may be a means by which greater value can be created in the South, such as in the remanufacturing of end-of-life products for re-export to customers in the North, e.g. Barloworld's refurbishment of Caterpillar parts in South Africa. The Green Alliance (Morgan and Mitchell, 2015) has postulated that CE will generate a higher number of jobs in the industrialised countries, but there is no strong evidence that this will be the case in Africa. As a result, the inequality that exists between the very poor without employment and those with permanent jobs may well continue in the CE.

Power relations and inequality

Power relations and institutional relationships impact the ability of actors in African countries to implement CE policies and business models. In the past, the GVCs of companies in the Global North have reflected an imbalance of power relations with producers in the South which favours buyers in wealthier economies. Schröder et al. (2018) highlight the power imbalances that exist in GVCs under a linear economy as transnational corporations and governments have greater access to resources and capital. Developing countries are now looking to create closed-loop value chains to reduce inequalities and generate fairer access to necessary material inputs. The result of these power imbalances is often inequality: income, gender and employment. Electronic and plastic waste are two examples where such inequalities are seen. People in low-income countries are engaging in waste recovery where local authorities have inefficient waste management services.

Noble Gonzalez and Schröder (2017) highlight Brazil and India, where informal waste picking is an essential source of employment for people, often in co-operatives, who are living on low incomes. In Africa there are many examples of rubbish sites where this is common, for example Olusosun, Lagos, Nigeria, and Agbogbloshie, Accra, Ghana. Yet, informal waste pickers could be one route for circular practices to develop in countries which have been unable to implement recycling of plastics and electronic waste at scale. In India, waste pickers undertake collecting, recycling and trading waste, which have contributed to the UN Sustainable Development Goals (SDGs) (WBCSD, 2016). This approach is now being seen in Africa, where small businesses and self-employed persons are working to generate an income with little legal recognition and low capital investment.

As emerging economies look to reduce inequalities, careful sourcing of minerals may be one route. This is seen in Fairphone's use of conflict-free materials in their modular smartphones (Fairphone, 2017). Similar criteria are used in the accreditation of gold mines in Africa for the extraction of the metal to the standards of the Fairtrade Foundation (Fairtrade Foundation, 2013).

In Africa, the issue of corruption can cause confusion in power relations. In 2017 the South African government closed the tyre recycler REDISA due to an unlawful misappropriation of public funds by REDISA directors (Business Report, 2017). Whilst the business model was demonstrating a positive outcome, close oversight of the company was lacking. Often, closed-loop initiatives take longer to implement and sometimes overlook issues of unequal power relations between producers (Schröder, Anggraeni and Weber, 2018).

The full impact of CE policies on the Global South is not yet fully understood, e.g. EU's CE Package, Eco-design Measures, and Extended Producer Responsibility (EPR) legislation. Furthermore, the negative impact generated by the Global North exporting waste for disposal to Africa has to be reconsidered from a health perspective as well as for its detrimental effect on the world's oceans.

Development issues that the circular economy can address

CE in Africa is focussed primarily on environmental and economic benefits. Yet potentially it could address the reduction of poverty and inequality. Alex Lemille (2016) has focussed on this element of CE and proposes that it can provide a route to supporting the human development of African countries through his CE 2.0 framework. This builds on the Ellen MacArthur Foundation's (2013a) 'Butterfly' diagram by including the 'humansphere' between the technical and biological elements. CE in Africa is often considered a concept which is only relevant to industrialised countries, but this viewpoint may limit the positive impact that circular approaches could have in creating employment and designing profitable business models.

In respect of poverty, Tearfund, a UK-based international development agency, is undertaking research into the opportunities for CE to improve livelihoods to assess the benefits of CE for developing countries and emerging economies (Fernandes, 2016). In Brazil the research (Fernandes, 2016) concluded that there are environmental benefits to circular approaches, there are opportunities to involve those at the base of the pyramid, productivity and employment can be increased, and there is a role for government to create the right enabling environment. Other emerging economy issues can be addressed by CE, such as waste management, disposal of electronic waste, energy needs (including the use of renewable energy), poverty reduction and job creation (Gower and Schröder, 2016).

Who is driving the transformation to a circular economy in Africa?

In the Global North, one of the key drivers for the transition to CE is the potential economic benefit. Recent research indicated a global US$ 4.5 trillion prize for turning what we currently waste into economic value by 2030 (Lacy and Rutqvist, 2015), such as the underutilisation of natural resources, products and assets. At a high level the authors saw three key drivers of the CE: resource constraints, technological development and socio-economic opportunity. They proposed five new business models which would provide a strong business case for CE: circular supply chain, recovery and recycling, product life extension, sharing platform, and product as service. Despite global supply chains engaging producers in the Global South, most of the case studies they include in the book are from the Global North. Cultural and contextual reasons may mean that application of these principles in developing countries might take longer than in Europe or North America.

Transitioning to CE in African countries will be a combination of effort by many actors, sometimes in coalition, but mostly in unstructured form. International organisations are emerging that support the sharing of good circular practices. One example is the Circular Economy Club (CEC, 2018); this is a global network of CE professionals which encourages collaboration to achieve a greater impact of circular practices. In February 2018 a global mapping session was held to record CE initiatives in 67 cities (including Cape Town and Port Harcourt), which has been shared publicly on an open source basis.

There are numerous actors in Africa building on the effort to create a more sustainable future for the continent: governments (e.g. The African Circular Economy Alliance – see text box), non-governmental organisations (e.g. The African Circular Economy Network – see text box), business (both multinational and smaller entrepreneurs), international development agencies (e.g. Tearfund) and international co-ordinating bodies (e.g. World Economic Forum).

BOX 9.1 CIRCULAR ECONOMY STAKEHOLDER NETWORKS IN AFRICA.

The African Circular Economy Alliance (ACEA) was announced at COP23 in Bonn in November 2017 as a collaboration between the governments of Rwanda, Nigeria and South Africa. 'We are looking at linking up the various projects and programmes on the continent and stimulating momentum towards the transformation to a CE', said Edna Molewa, Minister of Environmental Affairs, South Africa (Molewa, 2017). ACEA aims to encourage other African countries to build CE policies such as Nigeria's Extended Producers Responsibility Programme, which aims to ensure that businesses protect the environment and manage waste responsibly.

The African Circular Economy Network (ACEN, 2018) was formed in Cape Town in June 2016 by a group of CE professionals. The vision of ACEN is to build a restorative African economy that generates well-being and prosperity inclusive of all its people through new forms of economic production and consumption which maintain and regenerate its environmental resources (www.acen.africa). ACEN has initiated learning and knowledge sharing events between CE professionals in both developed and developing countries, so lessons learnt from the practical application of CE principles can be applied in different contexts.

Circular economy policies in Africa

Government policy in Africa has a major role to play at both the national and local levels. There is currently little CE-specific legislation, and so regulations and policies in operation are generally focussed on climate change mitigation, the Green Economy (GE), and waste management. Proposals are often presented but are still awaiting promulgation into government policy and legislation. In Ghana, the lack of an enabling environment, including financial and other incentives, is a major constraint for the creation of CE, particularly for entrepreneurs to set up informal repair businesses. The power exercised through the EU Extended Producer Responsibility (EPR) legislation by Northern manufacturers in the value chain will become a greater force for change and localised EPR legislation will be less important.

There are few systematic studies of CE policies in Africa, and so identification of policies currently relies on informal research approaches. Table 9.1 summarises some of CE-related policies that are in existence for a selection of African countries. Further research is required to identify the extent and impact of sustainability legislation and policies, such as waste management, recycling, extended producer responsibility, repair and renewable energy.

The examples given here are only a small selection of government policies currently in existence. The major challenge in the future will be to ensure that legislation is enforced. Furthermore, regulations behind more complex elements of CE, such as designing products as services and remanufacturing, are yet to be created at scale in Africa.

Examples of CE in Africa

In comparison to the shortage of CE policies and legislation in Africa, there are numerous examples of CE initiatives in Africa; some of these are summarised in Table 9.2. They have been categorised based on the Circle Economy 7 Key CE Principles framework (Circle Economy, 2018).

Kenya: CE case studies and policies

Whilst Kenya is yet to transform into a CE, it has made progress in transitioning to a Green Economy (GE). The transformation to a CE has to be seen in the context of the Kenyan government's GE strategy.

TABLE 9.1 CE-related policies, regulations and initiatives in a selection of African countries.

Country	Name of policy and year	Implementing agencies	Short description
Ethiopia	Climate Resilient Green Economy (2011)	Ministry of Environment	Reduce impact of climate change through renewable energy
Ghana	Ghana Goes for Green Growth (2010)	Ministry of Environment	Sustainable development and equitable low-carbon economic growth
Kenya	Nationally Appropriate Mitigation Action – Circular Economy Municipal Solid Waste Management Approach for Urban Areas (2016)	Ministry of Environment and Natural Resources	Diversion of waste from disposal sites towards recycling
Namibia	Green Economic Coalition Dialogue (2011)	Ministry of Labour and Social Welfare	Economic development, job creation and CO_2 reduction
Nigeria	Extended Producer Responsibility Programme (2013)	National Environmental Standards and Regulations Enforcement Agency	Minimisation of industrial waste and promotion of recycling
Rwanda	Plastic Bag Law 57 (2008)	Ministry of Natural Resources	Prohibition of manufacturing, importation, use and sale of polythene bags
South Africa	National Environmental Management Act (1998)	Ministry of Environment	Minimisation of waste, pollution and use of natural resources

Source: various Internet sources collected by Peter Desmond and Lara Maritano.

The Government of Kenya has taken several steps towards a GE and has developed a strategy that seeks to consolidate, scale up and embed green growth initiatives in national development goals. The Green Economy Strategy and Implementation Plan (GESIP) provides the overall policy framework to facilitate a transition to a GE and outlines the need to mainstream and align GE initiatives across the economic, social and environmental spheres (Gass, 2014). It is expected that a GE will protect the country's natural capital, reduce the environmental and

TABLE 9.2 Example CE case studies in Africa.

Circle Economy Principle	Initiative	Benefit
Prioritise regenerative sources	Biomimicry, South Africa	SPACE Project – water and waste treatment solutions in Langrug informal settlement using biomimicry principles to clean up the grey water, storm water, and solid waste challenges.
Design for the future	Mazzi Can, Uganda and Tanzania	Durable plastic to streamline the collection, storage and transport of milk from smallholder farmers.
Incorporate digital technology	COLIBA, Ivory Coast	Waste management mobile phone application in schools in Ghana, aiming to help users monetise their waste and satisfy the demand of recycling companies.
Collaborate to create joint value	Suame/Kumasi vehicle repair cluster, Ghana	Collective efficiency prolonging vehicle life achieved by 12,000 small workshops employing 200,00 workers.
	Government/World Bank, Tanzania,	Collaboration to develop more water-efficient practices among smallholder farmers.
Use waste as a resource	Sustainable Heating, South Africa, and EcoPost, Kenya	Biomass (residual carbon-based waste such as wood, sawdust, grain husks, sisal, etc.) is burned in furnaces. The heat from the combustion is used to feed local industries with steam, hot water, and heat. The by-product is ash, which farmers can use to enrich their soils.
	HP e-waste recycling, Kenya	Approximately 2,000 new jobs created within four years after launching the scheme.
Rethink the business model	Hello Tractor	Small-scale farmers request and pay for tractor services via SMS and mobile money.
Preserve and extend what's already made	Agbogbloshie, 'Old Fadama' slum, Accra, Ghana	Destination for locally generated automobile and electronic scrap collected from across the city of Accra.
	Imported e-waste, Nigeria	Approximately 70 per cent of imported waste is refurbished, tested and sold on.
Remanufacture	Barloworld, South Africa	Caterpillar parts repaired and refurbished with new guarantee.

Source: various Internet sources collected by Peter Desmond and Lara Maritano.

climate footprint, improve competitiveness, and spur economic growth, which in turn would create green and decent jobs.

The Green Economy Strategy is geared towards enabling Kenya to attain a higher economic growth rate consistent with the Vision 2030, which firmly embeds the principles of sustainable development in the overall national growth strategy. The policy framework for GE is designed to support a globally competitive low-carbon development path through promoting economic resilience and resource efficiency, sustainable management of natural resources, development of sustainable infrastructure, and providing support for social inclusion (Government of Kenya, 2016). It is spearheaded by the Ministry of Environment and Natural Resources through an inter-agency steering committee that consists of experts representing key government sectors, civil society and development partners.

In identifying the potential drivers for a GE, Klein et al. (2013) observe that the economic use of abundant natural resources could play a significant role. Effectively used and sustainably managed, these natural resources could offer great economic potential, which is currently under-exploited. Another important driver is the need for affordable and sustainable energy sources, as a great part of the population still lacks access to energy. Economic transformation to a CE and economic growth, necessary to lift people out of poverty, will require substantial energy resources. The high cost of connecting remote areas and communities to the grid make the development of decentralised sustainable energy solutions as well as the use of renewable energies important drivers on the way towards a GE. Kenya is beginning to address both the CE and GE through different approaches in the priority sectors of municipal waste management, waste electric and electronic equipment, and rural electrification.

Municipal waste management

Waste management is a major challenge in Kenya. The capital Nairobi produces around 2,400 tonnes of waste per day. Ninety-three per cent of this waste is potentially reusable, but only five per cent is recycled and composted and only 33 per cent is collected for disposal at Nairobi's single official dumpsite in Dandora. The rest is either disposed on illegal dumpsites, left next to houses or openly burnt. Both the official and illegal dumpsites are operated in an unsystematic, unplanned and highly unsanitary way. As a result, poorly managed solid waste pollutes the air, water and soil, causing significant health and environmental problems. This is especially true in slums and other low-income areas, where a high population density, paired with a lack of infrastructure and service provision, aggravates these problems (Soezer, 2016). This waste includes plastic shopping bags; according to UN Environment, over 100 million plastic shopping bags are given out every year by supermarkets in Kenya. About 4,000 tonnes of single-use plastic bags are produced each month and half of them end up in municipal waste streams that eventually block drainages and form a large portion of the garbage in major dumpsites (Kahinga, 2017).

Against this background, the Nationally Appropriate Mitigation Action (NAMA) on a Circular Economy Municipal Solid Waste Management Approach for Urban Areas was proposed. NAMA promotes an alternative to the existing waste value chain. Instead of waste being collected for disposal only, NAMA facilitates the diversion of 90 per cent of collected waste away from disposal sites and towards various recycling practices. The approach relates multiple links currently missing in the value chain: recycling points, where waste will be sorted for subsequent recycling, and composting facilities for the organic waste treatment. Under NAMA, up to 600 tonnes of waste will be recycled every day, which accounts for 25 per cent of Nairobi's total waste. This will save more than 800,000 tonnes in CO_2 emissions (over the 15 years' lifetime of NAMA) and add 1,600 jobs to the economy (Soezer, 2016).

The ban on some categories of plastics in the country was the first step towards establishing a well-functioning recycling sector that would open job opportunities (Kahinga, 2017). In addition, through many initiatives and with support from development partners, Kenya has made strides in mainstreaming the CE in small businesses in the country. It is estimated that a shift in investment to green sectors would lead to an additional 3.1 million people in Kenya being lifted out of poverty by 2030, while also contributing to a healthy GDP growth over the same period. Currently there are several small-scale traders and companies that are recycling and processing waste, which has created employment and is contributing to their livelihoods. They have teams that go round collecting the waste, which they later sort and recycle. The main challenge has been their limited processing capacity.

There are also some recycling companies that are turning plastic waste into building materials, furniture and other artefacts. EcoPost (www.ecopost.co.ke) is a social enterprise that addresses the challenges of urban waste management and plastic pollution, chronic youth unemployment, deforestation and climate change. EcoPost reported having processed 2.5 tonnes of plastic monthly and plans to triple that output. It has created 40 direct jobs and 5,000 indirect jobs. Over the next five years, it hopes to create at least 150 direct jobs and 20,000 indirect jobs.

Waste electrical and electronic equipment

Kenya has also witnessed exponential growth in the use of mobile phones, with the Communications Authority of Kenya (CAK) reporting that the number of users as of September 2017 stood at 41.0 million (CAK, 2017). This has mainly been driven by an influx of cheap short-life products, mainly from Asian countries. In 2016 for example, China's mobile phone exports to Kenya amounted to KSh 2.8 billion, according to data from the Kenya national Bureau of Statistics. Mobile handsets topped the list of items that local traders ordered from China, underlining the popularity of the low-priced smartphones. Due to the short lifecycles and a growing middle class, mobile phones are bought every 2–4 years, with the old or damaged handsets or their accessories being discarded as e-waste.

Waste of electrical and electronic equipment (WEEE), commonly referred to as e-waste, is the fastest growing waste component in Kenya as well as globally. UNEP estimates that over 17,000 tonnes of electronic waste are generated in Kenya annually. This is equivalent to 130 million mobile phones. The high rate of e-waste accumulation in Kenya is caused by short product lifecycles, the increasing affordability of electronics, and donations of used electronics from other countries. E-waste is composed of a complex mix of plastics and chemicals, including heavy metals and radioactive elements, which, when not properly handled, can be harmful to human health and the environment.

The WEEE Centre in Nairobi is one company that offers the service of awareness training and safe disposal of electrical and electronic waste in accordance with the National Environment Management Authority (NEMA) waste regulations and WEEE regulations, which are protective of both the environment and public health. The WEEE Centre processed approximately 200 tonnes of e-waste in 2016, which is only about 1 per cent of Kenya's total e-waste production. While plastic and metal can be processed in Kenya, other materials are deemed too hazardous and must be shipped abroad. Recycled e-waste is thus prepared before being shipped to Europe for processing.

To improve e-waste management, the Government of Kenya through the National Environment Management Authority (NEMA) has developed the draft Environmental Management and Co-ordination (E-Waste Management) Regulations 2013, which will provide an appropriate legal and institutional framework and mechanisms for the management of e-waste handling, collection, transportation, recycling and safe disposal of e-waste. It also provides for the improved legal and administrative co-ordination of the diverse sectoral initiatives in management of e-waste as a waste stream in order to improve the national capacity for the management of the e-waste (NEMA, 2017). If successful, the sustainable and safe management of e-waste can potentially turn hazardous work into green jobs through the collection, repair, recycling and processing of e-waste.

Rural electrification

Electricity in Kenya remains a preserve of a few, with national electrification levels remaining below 30 per cent on Kenya. The greatest challenge in effectively connecting rural communities to power grids has been the high price of an electricity connection. Currently, the price of a household connection is US$ 410, which is simply unaffordable for poor, rural households. An electricity connection costs approximately KSh 35,000 (Euro 318), and about Euro 0.1145 equivalent per kWh of electricity service. In addition, rural households in Kenya tend to be spread apart, and there are few straight lines through which one could easily run a power line. This makes it challenging for electricity planners to build cost-effective low voltage networks, particularly when they are unable to connect all of the neighbouring households at the same time (Lee, 2015).

These conditions present an opportunity for the development and adoption of renewable power generation technologies. Indeed, solar photovoltaic (PV) deployment has grown at unprecedented rates since the early 2000s. Global installed PV capacity reached 222 gigawatts (GW) at the end of 2015 and is expected to rise further to 4,500 GW by 2050. Particularly high cumulative deployment rates are expected by that time in China (1,731 GW), India (600 GW), the United States (600 GW), Japan (350 GW) and Germany (110 GW). As the global PV market increases, so will the volume of decommissioned PV panels. At the end of 2016, cumulative global PV waste streams are expected to have reached 43,500–250,000 metric tonnes. This is 0.1–0.6 per cent of the cumulative mass of all installed panels (4 million metric tonnes). Meanwhile, PV waste streams are bound to only increase further. Given an average panel lifetime of 30 years, large amounts of annual waste are anticipated by the early 2030s (IRENA and IEA-PVPS, 2016).

In 2014, new capacity additions of solar PV in Africa exceeded 800 MW, more than doubling the continent's cumulative installed PV capacity. This was followed by additions of 750 MW in 2015. There is also a fast-growing PV market to meet off-grid electricity needs. Technology improvements and lower costs have spurred local and social entrepreneurs in the solar home system (SHS) market and in developing stand-alone mini-grids. The years 2015 and 2016 have been recorded as breakthrough years for solar PV in Africa. However, ensuring that the right regulatory framework and institutional structures are in place will be an important pre-condition to maintain this growth. Efforts by development partners will therefore not only be essential from a capacity building perspective, but also potentially vital in using programmes to help de-risk solar PV projects (e.g. as is being done by the World Bank's Scaling Solar project). If these efforts are successful, the dividend for Africa could be significant. IRENA's REmap analysis – indicating a doubling of the share of renewable energy globally by 2030 – shows that Africa could be home to more than 70 GW of solar PV capacity by 2030 (Taylor, 2017).

In the wake of increasing solar power deployment in Africa, the continent is now battling the challenge of huge economic losses arising from a reliance on imported solar power equipment and tools, which are in high demand in Africa but cannot be made locally. While there are a few businesses involved in local production, varying from assembly of batteries and PV panels to simpler component manufacturing, the bulk of the solar products in use within the continent originates from foreign markets. Both funding as well as technical capacity for African solar projects are mostly sourced externally; primarily from Europe, America and Asia. Until power access is sufficiently improved to support robust local manufacturing in Africa, optimising solar power's potential and scaling its adoption will continue to rely on frameworks that support import trade pattern (Mama, 2018)

The benefits of solar traverse different aspects of consumer's lives. M-Kopa, a local solar energy provider, estimates that a customer saves about US$ 750 over the first four years by switching to its basic solar kit. The firm estimates that it has saved

US$ 206 million for the 300,000 households and businesses that use its products, while giving them 34 million hours of kerosene-free lighting a month. Solar energy eventually can provide clean, renewable power to millions of customers for whom affordable electricity has remained out of reach. This also benefits the environment while contributing positively to their customer's lives health-wise and economically, showing how solar lighting can start a virtuous cycle of development and progress (Faris, 2015). With about 46 per cent of the population living below the poverty line, the adoption of solar energy can combat poverty and provide new livelihoods opportunities.

The initiatives discussed, if implemented well, will impact positively on poverty in the country. High poverty levels result in a significant dependence on non-traded traditional biomass fuels due to the inaccessibility of other energy options, both financially and in terms of infrastructure development. As of 2007, traditional biomass energy resources in Kenya such as firewood, charcoal and agricultural wastes contributed approximately up to 70 per cent of Kenya's final energy demand and provided for almost 90 per cent of rural household energy needs; about one third in the form of charcoal and the rest from firewood.

South Africa: CE and mobile phones[1]

The experience in South Africa (SA) is set in a different economic, cultural and development context to that of Kenya. Mobile phones are quite ubiquitous in the modern world and the number of mobile phones in the world is understood to be greater than the number of traditional landlines. In 2014 there were estimated to be 4.5 billion mobile phone users in the world compared to 1.2 billion traditional fixed landlines (Mobiforge, 2015). Globally, this is generating significant amounts of under-utilised scarce resources, where significant numbers of phones are left in drawers when upgraded.

Data on the number of mobile phones sent to landfill globally each year is difficult to isolate as records are not kept by landfill operators to this level of detail. It is therefore not currently known how many mobile phones are sent to landfill or incineration each year in SA. Even though mobile phones form a small part of the e-waste stream in SA, they possess greater amounts of valuable materials in comparison to other types of e-waste, such as computers, photocopiers and printers. The constraints to adopting more circular approaches are wide-ranging and some relate to mobile phone design, manufacture, use and disposal processes. A co-ordinated effort by government, regulators, manufacturers, network operators, retailers and consumers is required to ensure that constraints to recycling are identified and strategies are developed to overcome them.

During the research, several solutions were identified. The key theme which emerged was the need to recognise the complexity of designing, manufacturing, and recycling mobile phones and the extent of the global value chains that result. The constraints identified from primary and secondary sources and examples of possible solutions are summarised in Table 9.3.

TABLE 9.3 Summary of constraints and circular economy solutions to mobile phones.

Constraint category	Secondary sources (literature)	Primary sources (interviews)	Example solution
Political	<ul><li>Lack of enabling environment and incentives</li><li>Poor implementation of policy</li><li>Manufacturer bearing the cost of disposal</li><li>Lack of recycling infrastructure</li><li>Informal sector not encouraged to engage with formal systems</li><li>Lack of co-ordinated national waste strategy</li></ul>	<ul><li>Definition of e-waste includes both electric and electronic</li><li>Lack of government support to formal and informal businesses involved in recycling</li><li>Manufacturers not being held to account for disposal costs</li><li>E-waste management is a complex process</li><li>No separation of recyclable materials in domestic waste collection</li><li>Silo working in government departments</li></ul>	*South African metropolitan city council partnering with European city to share learning of implementing recycling initiatives*
Economic	<ul><li>Poor return rate of phones for recycling</li><li>Lack of formal recycling schemes</li><li>Insufficient financial support from banks and other funders</li><li>Landfill is still a cheap option for disposal</li><li>Few incentives to develop CE initiatives</li></ul>	<ul><li>Lack of economically viable take-back schemes</li><li>Few financial incentives for consumers to return end-of-life devices</li><li>Informal economy does not have access to funding for parts</li><li>Raw material prices still too low to encourage reclaiming</li><li>Manufacturers focus on profit from selling new rather than second-hand phones</li><li>No current business case for investing in CE technologies</li><li>Informal economy not involved in collection and e-waste recycling</li></ul>	*Take-back scheme established in the UK through collaboration between retailer and environmental consultancy*
Social	<ul><li>Repair not seen as a viable option by consumers</li><li>Culture of retaining handsets in drawers</li><li>Collection systems not placed in convenient locations</li><li>Informal sector not suitably skilled to handle e-waste</li></ul>	<ul><li>Lack of trusted take-back scheme regarding personal privacy</li><li>End-of-use devices kept in drawers not available for recycling</li><li>Manufacturers not accepting EPR</li><li>Consumers still in ownership mindset rather than 'use'</li><li>Informal workers reluctant to register to participate in activities</li></ul>	*Creation of secure process to wipe confidential personal data from devices which have been returned under take-back schemes*

Constraint category	Secondary sources (literature)	Primary sources (interviews)	Example solution
Technological	• Mobile phones not designed to be repaired • Treatment of hazardous waste in limited locations • Lack of data on waste limits strategic planning • No government support for equipping informal sector with appropriate technology • Incompatibility of phone chargers	• Mobile phones not built to last and barriers to repair • Complexity of components hinders full recycling in SA • Lack of technical capability in SA for beneficiation work • Low tech approach to recycling missing out higher value components	*Informal sector working with other stakeholders to collect and process disposal of mobile phones in Ghana*
Legal	• Non-transferability of Western models to developing countries • Manufacturers resisted EPR legislation • Illegal importation of e-waste limited options for improving standards of recycling • No co-ordinated legislation for e-waste management	• Some of waste legislation is complex and lacks guidance on interpretation • Confusion over whether EPR legislation is enforceable • Lack of capability to draft and enforce e-waste regulations	*Extended Producer Responsibility (EPR) legislation has been successfully implemented in EU under the WEEE directive*
Environmental	• Poor control over informal workers polluting land with toxic materials • Air pollution from industrial processes limiting options	• Lack of local waste plans reduce efficiency of handling • Little separation of domestic recyclable waste • Phone design is non-environmentally friendly • No man-made alternatives to virgin raw materials in phone	*Lessons learnt from REDISA, tyre recycling initiative in South Africa, applied to mobile phone recycling*

Source: Desmond, 2016.

Manufacturers have a role to play in facilitating retailers and network operators to launch 'take-back' schemes to overcome the constraint of low return rates by consumers. Lessons can be learnt from successful CE applications in different sectors. An alternative 'servitisation' model would involve a consumer paying for the use of the phone (rather than owning it) and at the end of the contract the old handset would be taken back by the seller in return for a new phone and a new contract. This would replace the popular current system of free upgrades, which creates stocks of useable but unused handsets.

One element of CE for mobile phones involves the recycling of their component parts as well as reducing the amount of e-waste that ultimately goes to landfill and incineration, at the same time as recovering high-value metals, minerals and rare earths. Whilst the amount of these in each phone is small, the large volumes that can potentially be processed could make this commercially viable. Yet, given the amount of energy and materials contained within a smartphone, the most efficient circular approach is to keep the handset in use for the longest possible time through circular activities such as:

a   Reuse – a phone is passed onto another person to use with its original specification intact without major refurbishment from its original specification, often to friends and family.
b   Resell – the current market value of the phone is realised from a third party for cash or equivalent; these phones are either refurbished by the buyer or sent in bulk for auction.
c   Repair – when a device is broken and its useful life can be extended through the refurbishment and replacement of failed components.
d   Refurbish – new parts are added to the device to restore it to working order.
e   Remanufacture – the device is brought back to its original condition with some form of quality guarantee attached.

These interlinkage of options is summarised in Figure 9.1.

The ability to repair a phone requires manufacturers to design them so that components (such as camera, battery and Wi-Fi module) can be replaced or upgraded with assistance from online manuals such as ifixit.com. For example, consideration is now being given to the need for the redesign of mobile phones by the manufacturers using modular principles so that parts can be repaired and upgraded; the only currently commercially available phone is made by Fairphone (Fairphone, 2017).

One route to overcoming the constraints to efficiently recycling mobile phones in SA is the creation of a broader CE manifesto or roadmap on a country, regional or city basis, as has been created for a few countries and cities in the Global North, such as Finland (Pantsar et al., 2016) and London in the UK. A focus of possible future research is an evaluation of the EU Circular Economy Package (European Commission, 2017) and the BSI BS8001 CE standard (BSI, 2017) in an African context, particularly with regard to waste creation and resource utilisation. The role of the informal sector will need to become an additional dimension for South Africa as it has not have been covered in detail in CE roadmaps in European countries.

These constraints can be overcome through changes to legislation, policy, infrastructure, financial incentives and consumer behaviour, which would require all key stakeholders in collaboration participating in a globalised economy to recover valuable resources and reduce waste.

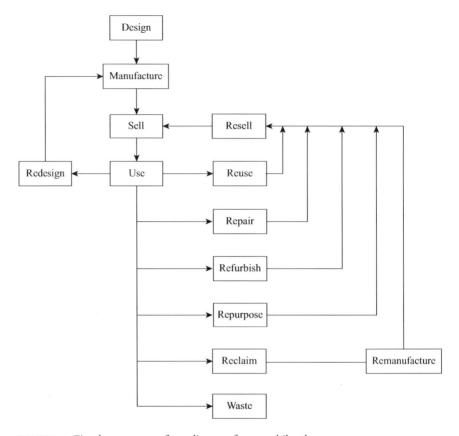

FIGURE 9.1 Circular economy flow diagram for a mobile phone.

Moving Africa towards a circular economy: conclusions and next steps

The transition towards CE in Africa can contribute towards the achievement of the SDGs. Gower and Schröder (2016) identify several as potential impact areas, particularly Goal 12 Sustainable Production and Consumption. The opportunities for accelerating CE principles in Africa are increasing and African case studies are emerging which reflect the local cultural context, such as Rwanda's new e-waste recycling facility which was opened in December 2017.

In Northern countries, the CE narrative is focussed on economic and environmental issues with less emphasis on social impacts. In comparison, CE activities in the Global South focus on social and environmental matters, with economic considerations only recently emerging. Circular practices have been in use in Africa for many years, yet case studies of good practice have remained largely hidden. This chapter has suggested that, whilst the transition towards CE in Africa will build on lessons learnt and principles developed in Northern countries, there will also be distinct differences in design, strategy and implementation of the CE in Africa.

Examples of current CE policies and practices have been considered in Kenya (renewable energy) and South Africa (e-waste) which, with others, can be shared throughout the continent. The main countries in Africa which are driving CE are Nigeria, South Africa, Rwanda and Kenya. Networks such as the African Circular Economy Alliance (development of national and local government policy) and the Africa Circular Economy Network (strategic application in business) working in collaboration will be able to facilitate this transition process. For Africa to transition from a linear to a circular economy, barriers need to be overcome through extensive collaboration between the various actors who each have a specific role to play.

Further research is required into how standards and CE policies decided in the Global North will be impacting African countries, and how African countries are implementing circular policies and business principles in their own context. The development of national, regional and city CE roadmaps, considering the views and experience of a wide variety of actors, will be an opportunity to co-ordinate action plans in individual countries. The potential benefits from a CE in Africa can then start to be realised with a positive impact on livelihoods, environment and job creation as well as working towards achieving the SDGs.

Note

1 This section is based on research for an MA dissertation (Desmond, 2016) into the constraints and solutions to recycling mobile phones in the Western Cape, South Africa.

References

ACEN (2018) About African Circular Economy Network. [Online] Available at: www.acen.africa/ [Accessed 11 August 2018].

British Standards Institution (2017) *BS 8001: 2017 framework for implementing the principles of the circular economy in organizations.* London: BSI.

Business Report (2017) Redisa waste tyre plan placed in final liquidation. Available at: www.iol.co.za/business-report/redisa-waste-tyre-plan-placed-in-final-liquidation-11397738 [Accessed 11 August 2018].

CAK (2017) *First Quarter Sector Statistics Report for the Financial Year 2017/2018.* Communications Authority of Kenya.

Circle Economy (2018) 7 key principles. [Online] Available at: www.circle-economy.com/the-7-key-elements-of-the-circular-economy [Accessed 31 May 2018].

CEC (2018) Circular Economy Club. [Online] Available at: https://circulareconomyclub.com [Accessed 12 August 2018].

Desmond, P. (2016) Towards a circular economy in South Africa: what are the constraints to recycling mobile phones? Master's dissertation. Institute of Development Studies.

Ellen MacArthur Foundation (2013a) *Towards the Circular Economy. Volume 1: economic and business rationale for an accelerated transition.* Isle of Wight: Ellen MacArthur Foundation.

Ellen MacArthur Foundation (2013b) *Towards the Circular Economy. Volume 2: opportunities for the consumer goods sector.* Isle of Wight: Ellen MacArthur Foundation.

European Commission (2017) *Report on the Implementation of the Circular Economy Action Plan*. Brussels: European Commission.
Fairphone (2017) Fairer materials: a list of the 10 we're focussing on. [Online] Available at: www.fairphone.com/en/2017/02/01/fairer-materials-a-list-of-the-next-10-were-taking-on/ [Accessed 31 May 2018].
Fairtrade Foundation (2013) New marks unveiled for fairtrade gold and precious metals. [Online] Available at: www.fairtrade.org.uk/Media-Centre/News/October-2013/New-marks-unveiled-for-Fairtrade-gold-and-precious-metals [Accessed 31 May 2018].
Faris, S. (2015) The solar company making a profit on poor Africans: M-Kopa plans to be a $1 billion company by selling solar panels to rural residents—and providing them with credit. [Online] Available at: www.bloomberg.com/features/2015-mkopa-solar-in-africa/ [Accessed on 10 April 2018].
Fernandes, A.G. (2016) *Closing the Loop: the benefits of the circular economy for developing countries and emerging economies*. London: Tearfund.
Gass, P. (2014) Kenya green economy strategy and implementation plan. [Online] Available at: www.iisd.org/project/kenya-green-economy-strategy-and-implementation-plan [Accessed 12 April 2018].
Government of Kenya (2016) *Green Economy Strategy and Implementation Plan 2016–2030: a low carbon, resource efficient, equitable and inclusive socio-economic transformation*. Nairobi: Government of Kenya.
Gower, R. and Schröder, P. (2016) *Virtuous Circle: how the circular economy can create jobs and save lives in low and middle-income countries*. London: Tearfund.
IRENA and IEA-PVPS (2016) *End-of-Life Management: solar photovoltaic panels*. Abu Dhabi: International Renewable Energy Agency and International Energy Agency Photovoltaic Power Systems. [Online] Available at: https://www.irena.org/publications/2016/Jun/End-of-life-management-Solar-Photovoltaic-Panels
IRP (2017) *Assessing Global Resource Use: a systems approach to resource efficiency and pollution reduction. A Report of the International Resource Panel*. Nairobi: UNEP.
Kahinga, E. (2017) Plastic bags and the circular economy in Kenya. [Online] Available at: www.kenyacic.org/news/plastic-bags-and-circular-economy-kenya [Accessed 15 April 2018].
Kirchherr, J., Reike, D. and Hekkert, M. (2017) *Conceptualizing the Circular Economy: an analysis of 114 definitions*. Utrecht: Innovation Studies Group, Copernicus Institute of Sustainable Development, Utrecht University.
Klein, J., Jochaud, P., Richter, H., Bechmann, R. and Hartmann, S. (2013) *Green Economy in Sub-Saharan Africa Lessons from Benin, Ethiopia, Ghana, Namibia and Nigeria*. Bonn: Deutsche Gesellschaft für internationale Zusammenarbeit Gmbh.
Lacy, P. and Rutqvist, J. (2015) *Waste to Wealth: creating advantage in a circular economy*. Basingstoke: Palgrave MacMillan.
Lee, K. (2015) Barriers to electrification for 'under grid' households in rural Kenya. National Bureau of Economic Research (NBER). [Online] Available at: http://blumcenter.berkeley.edu/news-posts/electrification-for-under-grid-households-in-rural-kenya-five-questions-for-ken-lee/ [Accessed on 12 April 2018].
Lemille, A. (2016) Circular economy 2.0. [Online] Available at: www.linkedin.com/pulse/circular-economy-20-alex-lemille [Accessed 31 May 2018].
Mama, C. (2018) Africa amidst 2018 global solar market predictions. [Online] Available at: https://solarmagazine.com [Accessed 27 May 2018].
Mobiforge (2015) Global mobile statistics 2014 part A: mobile subscribers; handset market share; mobile operators. [Online] Available at: https://mobiforge.com/researchanalysis/global-mobile-statistics-2014-part-a-mobile-subscribers-handset-market-share-mobileoperators [Accessed 12 August 2018].

Molewa, E. (2017) Minister Molewa's speech during launch of Africa Alliance on Circular Economy. Bonn, Germany, 11 November 2017. [Online] Available at: www.environment.gov.za/speech/molewa_cop23africaalliance_circular_economylaunch [Accessed 22 February 2018].

Morgan, J. and Mitchell, P. (2015) *Employment and the Circular Economy*. London: Green Alliance.

NEMA (2017) *Kenya Draft E- Waste Regulations*. National Environment Management Authority. [Online] Available at: www.nema.go.ke/index.php?option=com_content&view=article&id=35&Itemid=177 [Accessed 28 December 2017].

Noble Gonzalez, P. and Schröder, P. (2017) Closing the plastic waste loop: how do waste pickers contribute?Institute of Development Studies. [Online] Available at: www.ids.ac.uk/opinion/closing-the-plastic-waste-loop-how-do-waste-pickers-contribute [Accessed 5 May 2018].

Pantsar, M., Herlevi, K., Jarvinen, L. and Lalta, S. (2016) *Leading the Cycle: Finnish road map to a circular economy 2016–2025*. Helsinki: Sitra Studies.

Preston, F. and Lehne, J. (2017) *A Wider Circle? The circular economy in developing countries*. Energy, Environment and Resources Department Briefing. London: Chatham House.

Raworth, K. (2017) *Doughnut Economics: seven ways to think like a 21st-century economist*. New York: Random House.

SchmitzH. (2015) Africa's biggest recycling hub?IDS Blog. [Online] Available at: www.ids.ac.uk/opinion/africa-s-biggest-recycling-hub [Accessed 31 May 2018].

Schröder, P., Anggraeni, K. and Weber, U (2018) The relevance of circular economy practices to the sustainable development goals. *Journal of Industrial Ecology*. [Online] Available at: https://doi.org/10.1111/jiec.12732

Schröder, P., Dewick, P., Kusi-Sarpong, S. and Hofstetter, J.S. (2018) Circular economy and power relations in global value chains: tensions and trade-offs for lower income countries. *Resources, Conservation and Recycling*, 136, 77–78.

Soezer, A. (2016) A circular economy solid waste management approach for urban areas in Kenya. [Online] Available at: www.undp.org/content/ndc-support-programme/en/home/resources/namas/kenya-solid-waste-management-nama.html [Accessed 20 December 2017].

Taylor, M. (2017) Solar PV in Africa: falling costs driving rapid growth. [Online] Available at: https://www.esi-africa.com/solar-pv-africa-falling-costs-driving-rapid-growth [Accessed 15 May 2018].

Webster, K. (2015) *The Circular Economy: a wealth of flows*. Isle of Wight: Ellen MacArthur Foundation.

WBCSD (2016) *Informal Approaches Towards a Circular Economy*. World Business Council for Sustainable Development. [Online] Available at: www.wbcsd.org/Clusters/Circular-Economy-Factor10/Resources/Informal-approaches-towards-a-circular-economy [Accessed 18 May 2018].

PART IV
Livelihoods and traditional circular economy practices

10
SECURING NUTRITION THROUGH THE REVIVAL OF CIRCULAR LIFESTYLES

A case study of endogenous rural communities in Rajasthan

Deepak Sharma and Jayesh Joshi

Introduction

As per World Bank (2009) estimates, one-third of malnourished children in the world live in India. Furthermore, 53.1 per cent of Indian women (15 to 49 years old) are anaemic and 38.7 per cent of children below 5 years of age lag behind in growth rates (UNICEF, 2017). The National Family Health Survey (NFHS-4 2015–16) identifies under-nutrition as a leading problem in India, an observation that was evident to us in our work with endogenous farmers in North-Western India. Early in 2016, our interactions with tribal leaders, particularly with women members of self-help groups in the central-western tribal region of India, indicated to us that this situation of chronic under-nutrition had arisen due to the gradual disappearance of traditional farming systems that emphasised crop diversity. Our meetings brought us to the realisation that small and marginal farmers, who used to cultivate mainly for family nutrition, now had to prioritise growing food for the market. Market-controlled farming had resulted in a gradual shrinkage of food diversity associated with traditional diets, which were now at the verge of extinction.

Our organisation, VAAGDHARA is a civil society organisation that has been working with endogenous farmers for 20 years. We work with endogenous communities in rural India on issues of safe access to basic rights, water, food, nutrition, education, health services, livelihood improvement and sustainability. Our work includes study, action research, institution-building, awareness-generation, knowledge-building and providing sustained support to deprived communities in remote regions. Our focus group discussions with farmers in 2016 indicated to us that the linear approach of farming adopted post-green-revolution had made agriculture a high-input venture, and in this context, farmers were hard-pressed to sustain their livelihoods. Focus group participants were curious about the potential for gaining nutrition security through the cultivation and consumption of local foods (vegetable and animal based). This led us to the research question: can we can solve the problem of food insecurity by reinvigorating local diet diversity?

Our literature review revealed that small and marginal farmers within endogenous communities like the one we were working in are currently facing significant challenges, producing widespread suffering and increasing the rate of farmer suicides (Mishra, 2014). Thus, responding to community interest in local diet diversity and food security, VAAGDHARA started the groundwork to stimulate the revival of the traditional nutrition-sensitive farming system (NSFS). Several questions guided our efforts, including the following: how can NSFS be revived in a context where the present generation of farmers had limited practical knowledge about traditional nutritive food items and the processes of cultivating/collecting their nutritive values and prevent post-harvest losses. Information and evidence gaps existed, limiting community capacity to transition to a self-sustaining food system, thereby necessitating the design of interventions that could improve nutrition outcomes linked to traditional farming systems within this tribal belt of central India.

Transitions in agriculture and livelihoods in modern India

During pre-modern times, endogenous farming communities followed principles of circular economy in their farming, maintaining cycles of nutrient flow to ensure nutrition security for the family. Post the so-called 'green-revolution' in India, many farmers shifted their cultivation practices to mimic linear agricultural models. Over the decades, farming decisions have shifted from nutrition-focused family farming to market-led monocropping targeted for income generation. Modern market-oriented agriculture consists mostly of monoculture cropping that emphasises applying external inputs such as seeds, chemical fertiliser, pesticides and herbicides, along with the use of implements and technologies for harvesting. Such an approach focuses on exploiting natural resources (soil, water, energy) and utilising higher external inputs to get higher returns. These farming approaches are not suited to local environmental and cultural contexts. In the agricultural sector in India, resource-intense linear agricultural models have resulted in widespread food insecurity and malnutrition, in addition to being linked to environmental degradation, resource depletion, waste and pollution.

The linear economic model of agriculture as described by Allen (2015) and shown in Figure 10.1, reflects the 'take-make-throw away' principles of the dominant economic system (Schröder, 2018). This shift in farming from circular to linear models results in accelerated rates of nutrient depletion, causing a permanent loss of land, pollution of land and water bodies and eventually reducing the productivity of the land. It also results in a higher degree of nutrient export from a farmer's resources, i.e. their land, wells, ponds, pasture, and homestead, as well as common property resources. These ecosystem impacts have affected the

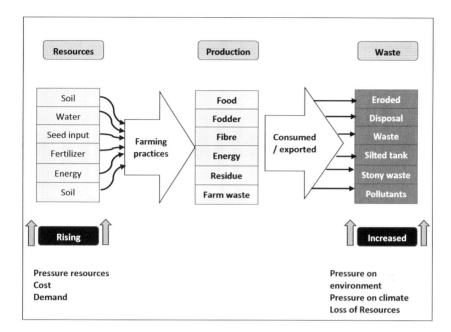

FIGURE 10.1 Linear model of agriculture.
Source: Adapted from Allen, 2015.

endogenous community in the Banswara region in three ways: first, the reduced food production for the family results in malnutrition. Second, market fluctuations lead to financial losses, pushing families further into a vicious cycle of poverty. Finally, the increased consumption of nutrients from soil produces wastelands and water pollution, leading to more scarcities and increasing vulnerabilities in the future.

Further consequences of ecosystem degradation manifest in the form of lower crop production, shortages of clean water, destroyed farmland and biodiversity losses (Rockström et al., 2009). Negative impacts on the immediate ecological system of farmers, food diversities and soil health are so high that the system's capacity to support family livelihoods is increasingly endangered. Recognising the urgency of this issue both in our immediate local context and globally, VAAGD-HARA conceptualised this research study around a 'Nutrition-Sensitive Farming System' (NSFS) and evolving intervention strategies and activities suitable for wider acceptance by communities, government and other development stakeholders, within the three-pillar perspective of LANSA[1] (Linking Agriculture and Nutrition in South Asia). The goals of the study were to design appropriate methods for promoting NSFS and to test them via participatory learning and action research, and through this evolve a framework for community-led NSFS to tackle undernutrition in tribal-dominated central India. Figure 10.2 shows how endogenous farmer communities in Banswara (VAAGDHARA), Orissa and Maharashtra

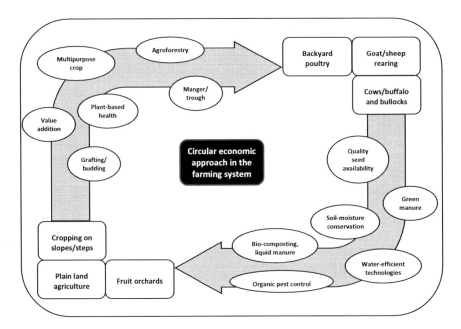

FIGURE 10.2 Circularity aspects of the revived farming system of the endogenous community.
Source: VAAGDHARA.

(MSSRF), Sundarbans (DRCRC) and other parts of India are reviving their farming system following circular economy approaches.

Circular economy and nutrition-sensitive farming systems

In the linear model of agriculture, farmers have increasingly had to go into debt to purchase the high-cost input resources necessary to sustain productivity. The escalating use of credit in agriculture is creating a milieu of exploitation. Furthermore, farmers do not always benefit from the fruits of their labour, as most of the production in terms of food, fodder, fibre, energy, residue and farm-waste is exported out of the system, and what is left behind are wastes in terms of eroded fields and silted tanks replete with pollutants. It was evident to us that the linear agricultural model was neither sustainable nor supportive of farmer livelihoods. In our search for alternatives, we turned to the circular economy (CE), hypothesising that the adoption of circular economy approaches in farming systems could provide a solution to this problem of increased eroded landscapes, silted tanks and polluted soils in agricultural lands.

According to the Ellen McArthur Foundation (EMF), the CE refers to a system that promotes restorative or regenerative process through deliberate product design. In this system, products and services are closed loops or 'cycles' that aim to retain as much value as possible for products, parts, and materials. This means that the aim should be to create a system that allows for long life and optimal reuse. The cyclic flows minimise losses of materials, resources and values, both qualitative and quantitative. The aim of a circular lifestyle is thus to minimise/eliminate the waste, utilise renewable sources of energy and phase out the use of harmful substances (Ellen MacArthur Foundation, 2013).

Taking up the CE concept in agriculture requires an analysis of the present high-input system which follows a more or less linear approach. This enables the identification of the potential opportunities and benefits a circular economy can yield for farming families, for the local environment and for the local economy. Within the circular economy vision, waste from a sub-system, for example cow dung from the animal husbandry sub-system, becomes an important input for agriculture, horticulture and pisciculture through proper composting and bio-digesting processes. Similarly, roughage from agriculture becomes fodder input for the livestock sub-system. Thus, implementing processes that emulate the circular economic approach can eventually reduce demand for raw materials. Furthermore, circular and closed-loop agriculture systems have direct benefits for biodiversity, soil conservation, the aquatic environment, and within the context of climate change (Harty, 2016).

Importantly, this circular economic model, as reflected in Figure 10.2, has its roots in farming systems of the past, where practices emphasised the prolonged circulation of resources in the farming system. Application of CE approaches in rural livelihoods can thus eliminate waste within the system. Adopting the principles of reduce, reuse and recycle pushes farmers to convert 'waste' produced in one part of the system into inputs and resources for another sub-system within the overall economy. The circular economy approach, particularly applied in agriculture, thus mimics the rhythms of the natural cycle of energy flow through different forms (biotic and abiotic), process (integration and disintegration), systems and strategies, which is in tune with the following three key principles, as outlined by the Ellen MacArthur Foundation (2016).

Principle 1: preserve and enhance natural capital by controlling finite stocks and balancing renewable resource flows

Principle 1 highlights the sustainability requisites of circular practices in terms of balancing resource utilisation and flows. Within agriculture, particularly in hilly areas, the first major challenge farmers face is preserving and enhancing the natural capital of soil, moisture, seeds and energy. Pursuing the linear model results in a loss of natural capital through soil erosion, fertility loss and lack of moisture. The use of non-renewable resources such as single-use seeds and fossil fuel energy to operate machinery produces both waste and greenhouse gas emissions. Thus, preserving and enhancing natural capital is a crucial principle to adopt while applying the circular economy lens to farming. This principle provides opportunities to assess the sustainability of different practices within farming. Our findings suggest that preserving and enhancing elements in soil, biodiversity and moisture within the system creates the conditions for regeneration of depleted natural capital.

Principle 2: optimise resource yields by circulating products, components and materials at the highest utility at all times in both technical and biological cycles

In farming systems, the application of Principle 2 connects the different sub-components of NSFS. In the linear model, animal husbandry and crop cultivation are two separate interventions carried out by two separate sectors. Thus, they demand two separate sets of inputs and produce two sets of outputs along with wastes. In contrast, circular practices provide scope for designing farm/production units that optimise resource yields by recycling materials and energy between different sub-systems to provide additional value. This can be done by creating material exchange loops, optimising multiple uses of input resources, sharing implements and exchanging outputs against inputs.

Principle 3: fostering system effectiveness by revealing and designing out negative externalities

Principle 3 focuses on reducing damage to ecosystems and livelihoods by identifying and eliminating negative externalities, which are aspects of the system that harm the overall efficiency and sustainability of the economy and associated livelihoods. Negative externalities could include food wastage, the release of pollutants into the soil and water, and continued dependence on single-use inputs. Designing out negative externalities requires learning from the system, and then managing externalities through proper land use, cover plantation, water management and avoiding pollutants. Optimising moisture use efficiency, for example, can be done by first identifying negative externalities (monocropping) and then eliminating it by shifting cultivation to mixed cropping. The tools applied to assess farming systems followed by a family or community could also serve to motivate others: data on the practices revived or adopted by the participants of the study groups can be used to demonstrate to other community members that these practices could be useful to them as well.

The methodology of participatory learning action (PLA) and data collection

Building on these theoretical ideas about the application of circular economy principles to agriculture, our study aimed to identify processes by which endogenous communities could attain a more sustainable life through the revival of circular approaches in farming. We explored the following questions: how can PLA tools be extended to revive the approach of the circular economy in nutrition-sensitive farming and address the problem of under-nutrition. The study evaluated the feasibility of utilising participatory learning activities to support the revival process, with the end goal being the establishment of a Nutrition-Sensitive Farming System for small and marginal farmers in Banswara. As these were new terms and concepts, we started with a literature review and formative interactions with community leaders, which was followed by an assessment of traditional and existing farming system through the lens of circularity.

Our study had the objective of developing and testing the applicability of participatory learning action (PLA) tools and then consolidating this learning in the form of a framework that could be used to facilitate the wider adoption of circular economy approaches in agriculture. PLA is an approach to learning that prioritises learning with and for communities (Jules et al., 1995). It combines an ever-growing toolkit of participatory and visual methods with natural interviewing techniques and is intended to facilitate a process of collective analysis and learning via the active participation of indigenous peoples and local communities (Jules et al., 1995; Swiderska et al., 2012). Based on our findings, we evolved a draft framework of PLA_NSFS that focused on the adoption of circular economy approaches by endogenous farmers to achieve nutrition security (Figure 10.3).

The PLA_NSFS includes ten learning sessions grouped into four separate phases, as shown in Figure 10.4, namely (1) triggering thought processes; (2) deciding and taking actions; (3) action monitoring; and (4) evaluating. Each of the ten sessions was normally 2–3 hours in duration and consisted of storytelling, pictorial displays, demonstrations, community actions and guided discussions.

As the PLA_NSFS process was extended for about ten months, facilitators were provided training in three phases. Each training lasted approximately two to three days and included content on three to four meetings under each phase. Before training facilitators, the implementing organisation was asked to collect information about food practices, nutrition status and the farming system in their communities, and were asked to identify resource agencies in their local area. This information was used to orient the facilitators to their local context. Facilitator review meetings were conducted on a fortnightly or monthly basis to provide ongoing support for field-related problems.

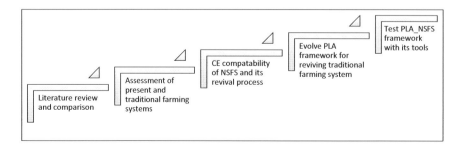

FIGURE 10.3 Framework and steps followed in the research programme.

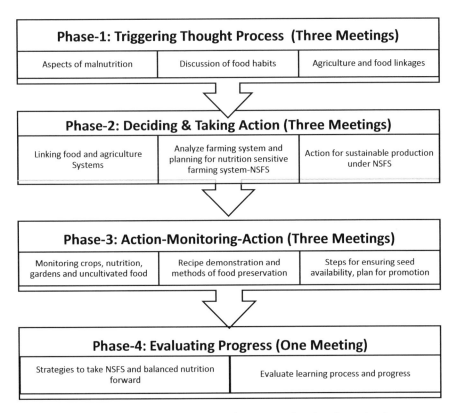

FIGURE 10.4 Framework of the participatory learning and action for a circular economy approach in NSFS.

Findings

The data generated in this study is from 30 groups in 30 villages of Banswara district. Each group comprised 20 women, with a total respondent number of 600 farmers. Group-wise data was collected and analysed to assess the effectiveness of various tools and sessions within PLA.[2] The analysis is also compared with findings from other regions, and the overall findings are presented in the following.

Phase 1: triggering thought processes

- The exercise focusing on 'What we eat vis-à-vis what we produce' was found to be an eye-opener for many farmers. The exercise unveiled the widening gap between cultivation and consumption patterns in a linear model of farming. This exercise was carried out in very first PLA session and served to trigger action and interest in the participants, who expressed an urge to deepen the links between their local farming and food systems

- The second session was dedicated to identifying the resource base each family utilised, and then linking access to resources with the farming (livelihood) system. This exercise served to trigger the identification of things families utilised regularly, but were not widely considered agricultural resources as such. For example, a plant named *buwari ghas* (broom grass) that is found in common lands along the roadside and termed a weed provided materials for brooms to one family for a whole year, at a value of US$ 5–6. If other families also used this plant to create brooms, it could provide US$ 500 to US$ 600 worth of value to the village. In this way, the community identified around 100 plant varieties to include within NSFS, many of which were not previously considered resource plants.
- Another important exercise focused on characterising the existing production systems of each family, the combined production systems of all 20 members as a group, and finally the production system of the entire village community. This exercise helped to establish the characteristics of each of these layers of community – family, hamlet, village, area etc. – and to understand whether these different production systems resemble linear, circular, mixed or hybrid economies. This exercise brought out the need for changes and laid the foundation for planning.
- Sessions on the basics of nutrition and health also helped build community understanding on the manifestations of under-nutrition, its causes and effects, and the methods of measuring nutrition status.

Phase 2: designing and taking action and Phase 3: action-monitoring-action

- Once the need for a shift in cultivation practices was established, the next exercise focused on designing farming practices with circular economy principles in mind. This exercise took more time than initially planned, as various non-farming tasks were identified as necessary at an earlier stage. One new action generated via the participatory process at the village level included creating a list of uncultivated/unconsumed food items: we listed around 36 food items which are found in the vicinity but not consumed by people in the present system because of various socio-economic issues. This exercise helped many people to recall the flowers, fruits and practices associated with food from their childhood. Groups also started looking for options to revive these foods and practices as well as find ways to transmit this knowledge to the next generation to ensure inter-generational knowledge transfer.
- The plans for adopting NSFS on the household level had various components, but the common component was developing a nutrition kitchen garden for each participating family. Therefore, another learning session on the whole cropping season included a review of farming practices as compared to planned processes.

- We found that endogenous communities have tremendous knowledge about a large number of uncultivated traditional food items and recipes. We plan to facilitate the adoption of these food practices by facilitating another stakeholder process dedicated to food and nutrition security among small and marginal farming families via the use of indigenous knowledge sources.
- Plans were prepared to improve farm diversity through nutrition gardens, integrated farming, and conserving and optimising village resources. PLA resulted in greater diversity and sustainable production in the farm with efficient nutrition practices.

Phase 3: evaluation and outcomes

- PLA demonstrated the link between environmental protection and sustainable natural resource management such as soil and water conservation, soil building, soil fertility maintenance, agroforestry, animal husbandry and cyclic flow of nutrients from different subsystems.
- The adoption of circular economy practices helped small and marginal farmers in moving towards sustainability by reducing input costs on different fronts, such as seeds, fertiliser, pest control, implements and even labour. The whole process could help 600 small and marginal tribal families re-establish nutrient flows and energy cycles at the farm level by using plants biomass (mostly fodder varieties) as food for animals and by recycling animal waste as compost to enhance soil nutrient content.
- Applying key principles of the circular economy showed the community that the traditional farming practices of the endogenous tribal community of the Banswara region used to follow more or less circular lifestyle concepts. Thus, it was possible to revive these practices by following the concepts of reduction (fodder saving, proper storage, seed selection), recycling (composting, organic waste composting, bio-composting) and reuse (adapting to multi-use implements, planters etc.)
- PLA_NSFS not only helped achieve family nutrition security but also augmented family income by 8–14 per cent through the sharing and sale of surplus production of some specific crops. Participating families earned a higher income as compared to what they used to earn when they produced crops based on market preferences.

In summary, our research demonstrated that NSFS reduces input, energy and opportunity costs, and in turn decreases resource dependence in agriculture. It results in improved nutrition outcomes, behavioural change and improved knowledge about food diversity and farming practices.

Conclusions and recommendations

Adoption of NSFS presents a win-win-win situation for small and marginal farmers by addressing the problem of malnutrition, supporting the family economy and maintaining the local environment. In the long run, this transformation could offer an essential competitive advantage to create more value from resources, lowering costs (both input and environmental), increasing production and securing food and nutrient supplies to the family. Adapting cyclic approaches in production and consumption within the farming system may significantly reduce resource extraction, alleviating the problem of resource scarcity. The concept of reduce-reuse-recycling, which is at the core of circular lifestyles and sustainable consumption and production, combined with organic agriculture can help in preventing agricultural pollution while contributing to a healthier society. This goes hand in hand with the recommendations of the Food and Agriculture Organisation (FAO, 2014), MSSRF (2018) and the LANSA community, who believe that agriculture for nutrition is to be prioritised over farming for the market.

Through participation in a PLA process, a community of 600 endogenous farmers in 30 villages learned to link nutrition with agricultural practices and the farming system. These families participated and contributed to the overall validity of the results. Executing ten tools with 30 groups allowed VAAGDHARA to evolve an implementation process in collaboration with the community. This collaborative process ensured the inclusion of diverse perspectives and broadened our understanding of how sustainable consumption and production can be revitalised in rural settings, particularly with tribal communities. Our findings pertaining to the PLA approach demonstrate that this can be an appropriate tool for promoting NSFS, even in contexts where farmers are facing urgent challenges related to environmental degradation, poverty and malnutrition. The PLA approach helps in identifying factors that either hinder or enable circularity within farming systems. Our research points to the promise of adapting the PLA learning cycle to help community groups adopt CE and NSFS, and through this overcome the vicious cycle of malnutrition and poverty.

Beyond participatory approaches, the role of government policies is also critical in deciding the fate of food systems, agricultural practices and farmer well-being. In India, government policy and investment have promoted wheat and rice, both largely produced using resource-intensive agricultural practices, as the major food grains. The distribution of rice and wheat via the public distribution system has influenced traditional farming practices. Thus, until national level policymaking on agriculture and food production shifts, local community efforts might come to naught. The implementation of PLA_NSFS will not in itself result in achieving SDG1 'No Poverty' or SDG2 'Zero Hunger'. This would first require that the NSFS lens be applied to the various policies of the government that influence the agriculture and food systems of endogenous communities. A future research question that we hope to examine on this issue is: 'what is the role of government policies in changing farming and food system of endogenous communities?' and 'is there competition between existing government policies vis-à-vis circular economy approaches for agricultural development?'

According to Davidova and Thomson (2014), 'family farming is often more than a professional occupation because it reflects a lifestyle based on beliefs and traditions about living and work', which goes hand in hand with NSFS approach promoted by this study. Our experiences in working with endogenous communities in rural India in reviving the traditions of CE suggest to us that this process could also enable a change in the power relationships in agriculture. The current linear model of agriculture exploits the most vulnerable small and marginal farmers, who are squeezed by a cycle of debt. Debt and failed crops result in widespread malnutrition, which affects women and children in the family disproportionately. A circular farming approach might be a way out of this situation of suffering. Access to nutritive food will improve the lives of women and children, potentially leading to changes in family power dynamics as well. Our experiences indicate the importance of applying the CE lens to agriculture and other allied occupations to alter power relations, bringing farmers sovereignty and ensuring nutrition security.

The VAAGDHARA experience demonstrates the potential for applying principles of the circular economy to one of the world's largest and oldest occupations: farming. Circular farming practices can help achieve the SDGs, in particular, SDG 2 'Zero Hunger', within the global agenda of sustainable consumption and production and address the problem of malnutrition through local solutions. VAAGDHARA plans to take this research agenda to diverse communities, collaborating with other stakeholders in the country, in the wider Asia region and beyond. This research also points to the need to evaluate public policy in agriculture with the lens of the CE and check the compatibility of policies, rules, and regulations with CE principles. In future work, VAAGDHARA also plans to evolve guidelines to define circularity within farming systems and generate an assessment of its social, environmental and economic benefits.

Acknowledgements

The paper is part of the research generated by the Leveraging Agriculture for Nutrition in South Asia (LANSA) research consortium funded by UK Aid from the UK Government. The views expressed do not necessarily reflect the UK government's official policies. The authors acknowledge the contribution of all respondent groups for sharing their knowledge about their rich heritage of farming systems and lifestyles, which go hand in hand with the concept of sustainable consumption and production through the diversity of indigenous farming and food system. They thank the team at LANSA for carrying forward the whole concept of sustainable consumption and production towards NSFS.

Notes

1. Consortium of research and development agencies in South Asia dedicated to linking agriculture and nutrition.
2. Process for each of the tools within the PLA approach is documented separately as part of the PLA_NSFS workbook (*in the Hindi language*) and can be accessed from VAAGDHARA by emailing jjoshi@vaagdhara.org.

References

Allen, B. (2015) The role of agriculture and forestry in circular economy. EIP-AGRI Workshop on Circular Economy, 28 October 2015. Naantali, Finland: Institute for European Environmental Policy (IEEP).

Davidova, S. and Thomson, K. (2014) *Family Farming in Europe: challenges and prospects*. Brussels: European Parliament.

Ellen MacArthur Foundation (2013) *Towards the Circular Economy*. Cowes: Ellen MacArthur Foundation.

Ellen MacArthur Foundation (2016) *Circular Economy in India: rethinking growth for long-term prosperity*. Cowes: Ellen MacArthur Foundation.

FAO (2014) *Family Farming: feeding the world, caring for the earth*. Rome: Food and Agriculture Organisation.

Harty, F. (2016) *Closed Loop Agriculture for Environmental Enhancement: returning biomass nutrients from humanure and urine to agriculture*. The Foundation for the Economics of Sustainability (Feasta). [Online] Available at: www.feasta.org

Jules, N., Guijit, I., Thompson, J. and Soones, I. (1995) *Participatory Learning and Action: a trainer's guide*. London: International Institute for Environment and Development.

LANSA (2017) *Farming Systems for Improved Nutrition: a formative study*. Working Paper Series. [Online] Available at: www.opendocs.ids.ac.uk

Mishra, S. (2014) *Farmers' Suicides in India, 1995–2012: measurement and interpretation*. London: Asia Research Centre, LSE. [Online] Available at: www.lse.ac.uk/asiaResearchCentre/_files/ARCWP62-Mishra.pdf

MSSRF (2018) *Farming System for Nutrition: a manual*. Chennai: M S Swaminathan Research Foundation. [Online] Available at: www.lansasouthasia.org

Rockström, J., Steffen, W., Noone, K., Persson, Å., Chapin, F.S., Lambin III, E., Lenton, T.M., Scheffer, M., Folke, C., Schellnhuber, H., Nykvist, B., De Wit, C.A., Hughes, T., van der Leeuw, S., Rodhe, H., Sörlin, S., Snyder, P.K., Costanza, R., Svedin, U., Falkenmark, M., Karlberg, L., Corell, R.W., Fabry, V.J., Hansen, J., Walker, B., Liverman, D., Richardson, K., Crutzen, P.and FoleyJ. (2009) Planetary boundaries: exploring the safe operating space for humanity. *Ecology and Society*, 14 (2), 32.

Schröder, P. (2018) *Can the Circular Economy Design Out Inequality as Well as Waste?*IDS Blog. Brighton: Institute of Development Studies.

Swiderska, K., Milligan, A., Kohli, K., Jonas, H., Shrumm, H., Hiemstra, W. and Oliva, M. (2012) *PLA 65 – Biodiversity and Culture: exploring community protocols, rights, and consent*. London: International Institute for Environment and Development.

UNICEF (2017) *Reducing Stunting in Children Under Five Years of Age: a comprehensive evaluation of UNICEF's strategies and programme performance – India country case study*. New York: United Nations Children's Fund.

World Bank (2009) *India's Undernourished Children: a call for reform and action*. Washington, DC: World Bank.

11
CONTESTING THOUGHTS AND ATTITUDES TO 'SUFFICIENCY'

Organic farming in an urbanised village in Thailand

Atsushi Watabe

Introduction

Rural transitions in Thailand

The Thai economy has experienced dramatic growth since the late 20th century. While it is the manufacturing and service sectors in urban areas that have been the main drivers of this rapid growth, the agricultural sector and rural areas have also seen remarkable development, even though the growth rate has been far below that of other sectors (Dixon, 1998). This economic development has had various impacts on rural societies: firstly, expansion of cultivated areas (Leturque and Wiggins, 2011) and imported agrochemicals (Greenpeace, 2008; Nguyen, 2016) were the driving factors in the growth of production focused on the export market. Additionally, contract farming expanded (Sriboonchitta and Wiboonpoongse, 2008). These characteristics indicate that Thailand's agriculture was developed as a 'take-make-use' linear economy depending on external inputs, not as a regenerative circular system. Secondly, rural populations have become integrated into the market economy. The expansion of commercial production in association with the spread of agrochemicals meant that farmers need to spend more capital on agricultural inputs and earn a greater income from their farm products. Moreover, farmers learned to grow a wide variety of commercial crops, carried out seasonal migration, and entered into the wage labour market in the industrial and commercial sectors developed in the late 20th century (De Koninck, Rigg and Vandergeest, 2012). Thirdly, many development projects have been introduced into rural areas to improve the socioeconomic conditions of the rural population, which had been left behind following industrial development. These policies included, but were not limited to, the microfinance scheme, the universal healthcare programme and community-scale income generation activities

(Watabe, 2016). Fourth, due to such changes, it may no longer be the size of one's landholding, but rather the capacity to induce and take advantage of development projects and off-farm employment that now has a decisive influence on one's social status in current rural societies in Thailand (Kelly, 2012; Walker, 2012).

Organic farming and its implications for the circular economy and the Sufficiency Economy

Whereas mainstream agriculture has been oriented to the export market, local organic farming has emerged recently. The Thai government launched the National Strategic Plan for Organic Agriculture Development (2008–2012) in 2008 and established the National Organic Agriculture Committee to oversee policy development in 2012. Several certification schemes co-exist, including both public and nongovernmental ones (Win, 2017). In Europe, organic farming is said to have potential contributions to the creation of a circular economy by supplying recyclable organic materials and preventing waste (EIP-AGRI, 2015; Longo, 2016; Dimitrov and Ivanova, 2017; Toop et al., 2017). We can expect similar effects from organic farming in Thailand on account of the past development of mainstream agriculture as a linear agricultural economy.

Besides, organic farming in Thailand is also expected to contribute to the resilience of farm households and rural society. Community development based on organic farming is the most popular model case for the practice of the philosophy of Sufficiency Economy, the official guiding principle of the national socioeconomic development. Sufficiency Economy gained wide attention following a speech by the late King Bhumibol Adulyadej (Rama IX) of Thailand on 4 December 1997. At that time, the shock of the Asian Financial Crisis caused the Thai people to lose confidence in the past model of rapid economic development, as it depended too much on external opportunities. His words touched the hearts of the people, prompting them to contemplate what kind of influence their behaviour had and to build self-immunity instead of pursuing profits. Following the speech, several national agencies initiated a discussion to elaborate on the philosophy and defined its three pillars, namely moderation (not too little and not too much), reasonableness (evaluating reasons for and the consequences of actions) and self-immunity (the ability to withstand shocks) as well as two supporting conditions: knowledge (wisdom) and integrity (virtue) (Ryratana, 2000; Office of the Royal Development Projects Board, 2004; United Nations Development Programme, 2007; Sangsuriyajan, 2011). The government has promoted this philosophy as the guiding principle of national development since its 9th National Economic and Social Development Plan and has registered hundreds of local practices as models of this philosophy (Office of the Royal Development Projects Board, 2004; Sufficiency Economy Movement Sub-committee, 2007; Chaiyawat, Piboolsravut and Kobsak Pootrakool, 2012). Many official documents highlight organic farming as a key measure for rural households and communities to build the three pillars of the Sufficiency Economy.

This philosophy also has received criticism, such as the fact that it presents an idealised picture of a self-sufficient rural society, which is far from the reality of either the traditional or modern Thai rural societies. Whereas rural villagers have sought better educational, employment and income opportunities (Walker, 2008, 2012; Rigg, Salamanca and Parnwell, 2012; Vandergeest, 2012; Wittayapak, 2012; Watabe, 2016), the emphasis on harmony or mutual support within the village communities may blur the existing uneven power relationships and inequalities (Walker, 2008). The method used in selecting these models has also been criticised. The government showcased many practices that resulted in success in turning the philosophy into reality. However, most of these cases were initiated or progressed irrespective of the philosophy, despite being subsequently registered as being based on the model. They were, in other words, a backdated application of the label of Sufficiency Economy (Singsuriya, 2017).

Organic farming may be effective for building a circular economy by making full use of organic resources and reducing imported chemicals, as well as enabling the Sufficiency Economy by helping communities increase resilience and prepare for external shocks. Since both concepts and approaches have some important similarities, it is no wonder that organic farming can contribute to the achievements of the two different ideas for sustainable development. The philosophy of Sufficiency Economy suggests that people pay attention to the impact of their behaviour on society and the environment, and protect themselves from external shocks. To this end, it encourages making the best use of the resources available in one's immediate circle. At least four of the six principles in the ReSOLVE framework of the circular economy, namely regenerate, share, optimise and loop (Ellen MacArthur Foundation, 2012; McKinsey and Company, 2016), also contribute to sufficiency. In fact, SWITCH-Asia's report on the circular economy in Asia recognised the influence of the Sufficiency Economy in promoting policies and practices aligned with Circular Economy (Satori, 2016). Meanwhile, it is important to also highlight a considerable difference between the two concepts: the Sufficiency Economy is considerably different from the Circular Economy in that it does not attach weight to market competitiveness.

Sustainability in a day-to-day context

Those who practice activities that are in line with Sufficiency Economy or circular economy concepts often do not do so for the sake of sustainability as encouraged by the philosophy of Sufficiency Economy or circular economy. As was noted earlier, people's economic conditions, as well as their social status in villages today, are not determined solely by the size of their landholding. Rural people are urged to reconfigure the resources and capacities they can rely on to seize opportunities for maintaining and improving their livelihoods. Sometimes people's day-to-day choices happen to coincide with the practices advocated by the Sufficiency Economy or circular economy, despite not being convinced of their principles. However, people sometimes make choices that are deemed unsustainable by experts or policymakers, even though they try to make the best use of the available resources to sustain their living.

Against this background, even among people living in the same areas, their framework to interpret the past and expect the future pathways are diverse and sometimes contesting (Leach, Scoones and Stirling, 2007; Stirling, Leach and Scoones, 2010), such that they respond differently when taking up opportunities and coping with the uncertainties. Thus, it is important to learn the meanings people attach to the various options of activities. including both 'sustainable' ones and less 'sustainable' ones, instead of narrowing down the focus on the 'sustainable' features of some of their activities.

In consideration of this, this chapter introduces a peri-urban village in Northeast Thailand. While the majority of households engage in off-farm employment, several villagers launched an organic farming initiative; later, the government labelled the entire village as a model of Sufficiency Economy. However, villagers have different attitudes regarding engaging in or distancing themselves from organic farming, which indicates that people live under varying conditions, and they also have diverse concerns and aspirations regarding their future. Thus, this case provides a clue to the way in which people interpret the ideas of sustainability such as Sufficiency Economy in the contexts of their everyday lives. It also enables us to look into how people manage their options when they have contesting ideas about the necessities and opportunities for sustaining their living.

An urbanised village in Northeast Thailand

D Village is located approximately 20 kilometres west of the centre of Khon Kaen city, one of the regional centres of Northeast Thailand. Due to the poor conditions of the soil and limited rainfall, the productivity of rice farming as well as other crops is limited. Thus, it has been difficult for many villagers to make a living only through agriculture on their farmland. However, the villagers are more affluent than the average villages in the Northeast.[1] Thanks to its location (alongside the highway), the villagers can commute to the city centre in 30 minutes. Since the 1980s, many villagers have also travelled abroad and worked at construction sites or factories in the Middle East and East Asia. In the 1990s, many factories commenced operation and employed hundreds of workers from the surrounding villages. As a result, villagers were able to access a large cash income. In the early 2000s, most villagers had built concrete houses and acquired a variety of consumer durables such electric household appliances and cars, identical to those possessed by families living in the city centre. Educational attainment was also better than that of other villages. Almost all children attended upper secondary school and many of the graduates are engaged in paid employment. One of the villagers told me in 2005 that 'the age of rice farming has finished'.

Villagers stated that they are responsible for providing their children with opportunities to 'study to as a high a level as possible, and to have as good a job as possible (*rian sun suun, heat ngaan di-dii*)'. Since the 1980s, this has been a common reason for villagers to seek every opportunity to diversify their income sources, such as planting new cash crops, establishing off-farm businesses like commodity stores, and migrating overseas for work. Consequently, many villagers were fortunate enough to assist their children in securing employment in factories, service or commercial sectors. Interestingly, however, this did not automatically lead to the dismantling of the farm households nor to the full-scale de-agrarianisation of the rural areas. The majority of the families have young members who commute to the cities or factories or who send money from varying provinces. They manage their households' budgets by combining several income sources, including the salaries of these members with the relatively small sales of their farm products.

During my visits to the village in April to May 2016 and May 2017, I interviewed the headmen of the hamlets and 12 villagers. These interviews revealed the changes experienced in the village during the previous decade, including the introduction of organic farming in 2005.[2]

It is interesting to note that an organic agriculture group was formed and subsequently became a national model of sufficiency in this kind of peri-urban village, where people tried to avoid the hardships of farming as a way of living and instead took advantage of the abundant off-farm opportunities. More noteworthy are the different attitudes among villagers to organic agriculture. While group members enthusiastically practice it, the majority of the villagers simply reduced chemical inputs without aiming to sell their 'organic' produce. In addition, some farmers continued their conventional practice of farming using chemicals.

Different responses to Sufficiency Economy and organic farming

This section examines three different groups in the village and their responses to organic farming activities. It considers the different conditions and socio-economic circumstances, such as family structures, assets and economic opportunities. These different conditions result in different mental frames to interpret past changes as well as future concerns and aspirations in relation to people's way of life.

Group 1: the organic farming group

Group activities

Organic farming spread within a few years among the remaining village farmers. The head of one of the two hamlets of D Village learned about the methods from a model village in another province and introduced these techniques to his village in 2005. Out of more than 1,400 villagers, 50 responded to his call to learn the organic farming techniques and formed a group at their experimental farm. The group's activities expanded over the next 12 years despite the fact that about 20 of the original members left the group. The main activities as of 2017 are summarised as follows:

- Production and sales of farm produce, such as eggplants, cucumbers, watermelons and herbs.
- Research and experimentation into new methodologies, including the improvement of soil quality, organic fertilisers, pest control and marketing.
- Production and sales of organic fertilisers at their learning centre.
- Frequently held lectures on farming methods to visitors at the learning centre. Core members are often invited to other villages, schools and local governments to give presentations on their experiences and methods.

The members usually work in the fields of individual families and occasionally gather at the experimental farm on a two *rai* (0.3 hectares) area of land dedicated by one of the members to carry out these activities. In 2009, the village was awarded the title of 'Sufficiency Economy Model Village for ASEAN' by the national government. The village received a subsidy and established a Sufficiency Economy Learning Centre at their formerly experimental farm. Due to this recognition from the authorities, D Village's organic project attained broad recognition. Many visitors from all over Thailand came to the village to buy vegetables or learn their methods. However, becoming famous can bring risks. Labels of fake 'D Village products' began to appear on the urban market in 2015. In response, the village decided not to sell their produce outside of the village, but rather asked buyers to come to their centre to protect the village's reputation.

The group's background

The organic farming group is mostly made up of older villagers. While other families have increased their cash income, these villagers do not have young family members who work in factories or in cities. The headman's objective when he initiated the group was to provide more livelihood options for those families left behind. The members of this group experienced difficulties participating in the labour market, or in other words, they failed to gain the benefit of the linear model of development for the region, so their activity was a reasonable way to make better use of what they had in order to protect their livelihoods.

However, organic farming turned out to bring several additional benefits to the group members. Firstly, it enabled them to reduce the financial costs of farming. The headman's family had previously spent THB 30,000 annually on growing rice and cassava, which included buying chemical fertilisers and pesticides. The cost has now been reduced to THB 6,000 per year. Secondly, members emphasise that their own health and that of their families has improved since they stopped using chemicals.

The methods used by the villagers enable them to make the best use of their land, labour and other resources available within their environment, reducing their dependence on imported products. Their activities also enabled them to gain more income from external markets. Their products sell for 10 to 20 per cent more than those grown with chemicals. Crop yields have increased due to the improved soil quality. The villagers also have an increased number of diversified income sources, including the sales of organic fertilisers and the honorarium they receive as lecturers. Thus, it is important to them not to restrict their activity to the village, but to properly control it so that they can connect to the outside world and reduce vulnerabilities.

Group members' considerations

Although the group members mostly do not have young family members who send remittances back home, they do rely on diverse income sources. To take the headman as an example, he earns tens of thousands of Thai baht in lecture fees, and THB 96,000 annually as a headman, in addition to the THB 100,000–200,000 from the sales of organic vegetables. In fact, even those who diligently practice 'sufficiency' organic farming understand that rural households survive by pursuing a combination of opportunities both within and outside the village. Thus, it is not surprising that group members acknowledge that organic farming can be an option for those unemployed. One of the group members stated:

> If one has nothing to do but farming and caring for one's health, then this is one option. Moreover, children may not be able to work in factories forever.
> *(An elderly woman, organic farming group member, 5 May 2017)*

Her statement is also noteworthy for the fact that she brings into the view the possibility that one's young family members might lose their factory jobs and become unemployed in the future.

The group members have a unique manner of talking about their activities, referring to the King's Sufficiency Economy philosophy.

> For many decades, His Majesty had taught us that we should refrain from making everything business, and should keep doing what we can do. It took time for us to understand what he said. But I am happy that villagers finally listened.
> *(The headman, organic farming group leader, 3 May 2016)*

> I am confident that our wellbeing (*khunapaap*) has improved. People now understand what is meant by happiness suggested by Sufficiency Economy. When we have money ourselves, we don't need to beg from others. Our health (*sukhapaap*) has also improved.
> *(The headman, organic farming group leader, 4 May 2017)*

These statements sound like formulaic phrases often used in the headman's lectures. These words are also interesting as they emphasise that the group was established out of their individual and community concerns, irrespective of the philosophy.

Group 2: families who stopped using agrochemicals

Although D Village became famous for organic farming, we should not forget that the group comprises only 30 villagers from a total of more than 1,400. To better understand what organic farming brought to the villager's lives, we should turn our attention away from the group of 30 members and see what the other villagers are doing. The mental frames that they have to interpret the past and present transitions as well as the future opportunities and threats are considerably different from those of the organic farming group members.

Labour-saving methods

The majority of the other farmers – most of whom were above 50 years old – did not join the group, but accepted the headman's call and stopped using chemical fertilisers and pesticides. They grow rice or vegetables mainly for personal consumption within their households. They do not need to increase their harvest for sale since their primary income sources are their children's salaries. This category of people partially adopted organic farming to reduce labour input.

As younger families work outside the village, they hire labourers and machines to continue farming with fewer hands and have (re-)introduced several practices to secure the workforce further and reduce labour. Firstly, crop sharing – once a common agreement between parents and children before inheriting land – is now agreed upon between elderly farmers and their neighbours without farmland. Previously, both the landowners (parents) and tenants (children) took 50 per cent of the harvest. This ratio has now changed to 25 to 30 per cent for the owners and 70 to 75 per cent for the sharecroppers, giving more benefits to the latter. Secondly, they have changed their farming methods. Rice seedlings have been planted since the late 20th century for productivity; however, a return has been made to the broadcast seeding method. Finally, they stopped using chemicals to save time needed for spraying and to avoid purchasing costs. All of these methods enable families to save on labour, but lead to a substantial reduction in yields. Thus, the spread of these methods clearly indicates that these families do not practice farming as their main income source. For this category of families, the elimination of chemical input forms part of the various interventions through which they continue their diversified livelihoods, in combination with other employment of their younger family members.

Considerations of the non-members who stopped using chemicals

In 2000, village farmlands were largely maintained by old villagers. They often said that they 'just do' rice and crop fields simply because they have the land. With such statements, they highlight the contrast of their farming activities and the profitable activities of their children and neighbours, whose salaries were essential to purchase daily necessities.

However, people now recognise that factories do not employ workers aged over 40–45 years. They must, therefore, consider a future in which their children are unable to continue to work in factories. They do not know if their children will return after their retirement. Even if they come back to the village, they may not go to work in the fields. However, the farmland provides options for their children. In this way, meaning is given to the farm activities that the older villagers perform at minimum cost and no profit, rather than 'just do(ing)' it because they have land. These farmers may also put more emphasis on participating in the organic farming group if they cannot depend on their children's salaries any longer. One elderly woman stated:

> I am not interested in it now, as I am not short of money. But I would consider it in cases when I am concerned (about money). For example, if my children are unemployed.
>
> *(An elderly woman, non-member, 5 May 2017)*

As was mentioned by the organic farming group members, one of the concerns among the old villagers is a future time when youths cannot work in factories or take up employment in the city. In light of this, the reduced use of chemicals as well as other methods of reducing costs are derived from their demand to maintain their current forms of livelihood and offer their children an increased number of options when they retire from factory work.

Group 3: villagers who still use chemicals

Multiple economic activities

Some of the other villagers still use chemicals. These families have younger members earning an income outside of the village. Additionally, they hold farmland of 15 *rai* (2.5 hectares) or larger, while an average household in D Village holds less than 6 *rai* (1 hectare). They are the wealthier families in D Village both regarding their net income and their assets. Although they have cash income from wage labour or business, they still work on making profits from selling commercial crops. Since they have younger families earning an income outside of the village, they do not have enough time to practice organic farming. Thus, it is natural that they stick to the conventional way of growing commercial crops such as corn, cassava and sugar on farmlands that are larger than those owned by an average household, and that they use a fair amount of chemicals.

The 'farmers' way represented by those who still use chemicals

In fact, these farmers say that it is imperative to use chemicals for those who subsist through farming. The term 'those who subsist through farming' sounds strange, as they largely obtain their income from salaries of family members working in factories or the city. However, they insist that they are the ones who continue what the farmers have been doing in the past. For several decades, the use of chemical fertilisers and pesticides has been one of the key agricultural developments that enabled them to enjoy a better quality of living, along with their efforts to introduce more profitable crops. The promotion of non-chemical farming, therefore, is not compatible with their former methods. Some of the farmers who hold higher educational backgrounds have questioned the way of thinking of the organic farming group members.

> They don't seem to be in their right mind when they say farmers can do without chemicals. We have depended on chemicals for decades. Can they go organic once they refrained from it for several years? It can't be that easy. Villagers do not understand such an obvious fact.
> *(Former NGO officer, non-member, 5 May 2017)*

They also note that only a fraction of people can afford to take on the enormous amount of work required to profit from organic farming. To such critics, organic farming is simply unrealistic. Such opinions show the contempt of the intellectual class for the villagers. However, such words reveal the importance of efforts taken to ensure the means of living for farmers, rather than pursuing idealistic practices of organic or sufficiency. This perspective partially corresponds to the original intention of the leaders of the organic farming group, namely to develop a method to be used by those who cannot currently depend on the most popular method to support their families.

Discussion and conclusion: different responses, different aspirations

We have seen three differing responses and narratives from the villagers with regards to organic farming, originating from the different conditions that influence their livelihoods means, in particular their access to the labour market and landholding. The varying responses towards organic farming as a regenerative circular agricultural method reflect the different mental frames of farmers interpreting the past and present conditions, prospects and aspirations of changing and maintaining their means of living.

The organic farming group's activity to make the most of what is at hand supports those who lag behind in socio-economic development, since they have limited access or capacity to make more from something brought in from outside or selling to external markets. In this sense, their model offers more than the growth strategies used by large companies or national economies. However, the group generates a substantial portion of their profits from selling their knowledge and products externally. It is important for them to control what is circulated by which means and what is brought in from and sent out to the outside world. We should recall the fact that it is unrealistic to become entirely self-sufficient, even for rural households living on the so-called 'Sufficiency' living model. On the other hand, the majority of the villagers rely on wages earned in factories, and also depend on farms using various cost-cutting measures. They adopted these methods to 'just' continue farming without profit, but later found that their farmland can provide additional options for their children when they retire from the labour market. Some of the families who still use chemicals pursue multiple wage opportunities in the wider agricultural markets. They emphasise that this has been the normal way for farmers to survive for decades. In this way, villagers carefully coordinate what they have at hand and what they must procure from outside. Those who do not adopt the 'Sufficiency' model also utilise both internal and external resources, which have been imperative for their efforts in giving their children the opportunities to live differently from previous generations. Thus, villagers have adopted these different responses as ways of maintaining their current livelihoods and securing future options at a time when they are not always able to rely on the assets and resources that they have had for decades.

It is also important to note that people do not necessarily stick to one of the three contesting responses described here. As was mentioned previously, about 20 of the 50 original members of the organic farming group have already shifted to cost-saving methods of farming. Some of the majority of villagers who 'just do' farming also revealed that they might consider joining the group in future. Those farmers who continue practising conventional chemical agriculture may not turn to organic farming, but they may also give up commercial farming and adopt cost-effective methods including 'non-chemical' when they become older. Thus, this case illustrates the multi-layered nature of the pathways for sustainable lifestyles and livelihoods. Even though it might seem that they have contesting ideas and options for maintaining and improving their living, they can make necessary adjustments to the different approaches for a more sustainable way of living.

This chapter looked at the socio-economic context of 'sustainable' practices, both for those who adopt such practices and those who do not, rather than narrowing our scope to the attractive features that correspond with sustainability concepts. The analysis revealed that in spite of the different attitudes to the 'sustainable' practices contributing to the building of a circular economy and Sufficiency Economy, all of these attitudes derive from their unique interpretations of a change in circumstances. It also showed that these different attitudes and the contesting ideas about sustainable living are not entirely incompatible, allowing people to adjust to different options when necessary. The pathways of transitions in a local society are made through such flexible adaptations by the people involved, rather than through the spread of a single concept of 'Sufficiency'.

Notes

1 The average annual income per person was recorded in 2017 as THB 130,000. This was highest among the province's rural areas.
2 Among the various certificates for organic or non-chemical agriculture, the group's activity is certified by Thai GAP (Good Agricultural Practice). Since villagers do not differentiate between 'organic agriculture (*Kaset Insii*)' and 'non-chemical agriculture (*Kaset Proat San Pit*)' in their conversation, this paper mainly uses the term 'organic' unless there are specific reasons not to.

References

Chaiyawat, W., Piboolsravut, P. and Pootrakool, K. (2012) *Sufficiency Economy Philosophy and Development*. Bangkok: The Crown Property Bureau. [Online] Available at: http://tica.thaigov.net/main/contents/files/business-20160904-174653-791776.pdf

Dimitrov, D.K. and Ivanova, M. (2017) Trends in organic farming development in Bulgaria: applying circular economy principles to sustainable rural development. *Visegrad Journal on Bioeconomy and Sustainable Development*, 6 (1), 10–16. [Online] Available at: http://doi.org/ 10.1515/vjbsd-2017-0002

Dixon, C. (1998) *The Thai Economy*. London and New York: Routledge.

EIP-AGRI (2015) *EIP-AGRI Workshop Opportunities for Agriculture and Forestry in the Circular Economy*. Workshop Report. [Online] Available at: http://ec.europa.eu/eip/agriculture/en/content/eip-agri-workshop-opportunities-agriculture-and-forestry-circular-economy-final-report

Ellen MacArthur Foundation (2012) *Towards a Circular Economy: economic and business rationale for an accelerated transition, greener management international*. Cowes: Ellen MacArthur Foundation.

Greenpeace (2008) Agrochemicals unmasked: fertilizer and pesticide use in Thailand and its consequences to the environment. [Online] Available at: www.greenpeace.org/seasia/th/press/reports/agrochemicals-in-thailand-eng/

Kelly, P.F. (2012) Class reproduction in a transitional agrarian setting: youth trajectories in a peri-urban Philippine village. In Rigg, J. and Vandergeest, P. (eds.) *Revisiting Rural Places Pathways to Poverty and Prosperity in Southeast Asia*, pp. 229–249. Singapore: Nus Press.

De Koninck, R., Rigg, J. and Vandergeest, P. (2012) A half century of agrarian transformation in Southeast Asia, 1960–2010. In Rigg, J. and Vandergeest, P. (eds.) *Revisiting Rural Places Pathways to Poverty and Prosperity in Southeast Asia*, pp. 25–37. Singapore: Nus Press.

Leach, M., Scoones, I. and Stirling, A. (2007) *Pathways to Sustainability: an overview of the STEPS Centre approach*. STEPS Approach Paper. Brighton: STEPS Centre. [Online] Available at: http://sro.sussex.ac.uk/26103/

Leturque, H. and Wiggins, S. (2011) *Thailand's Progress in Agriculture: transition and sustained productivity growth*. London: Overseas Development Institute.

Longo, A. (2016) *Circular Economy and Agriculture*. Brussels: European Landowners' Organization. [Online] Available at: www.europeanlandowners.org/files/Intergroup/2016/Conference on CE agriculture final.pdf

McKinsey and Company (2016) The circular economy: moving from theory to practice. McKinsey Center for Business and Environment. Special Edition. [Online] Available at: www.mckinsey.com/~/media/McKinsey/Business Functions/Sustainability and Resource Productivity/Our Insights/The circular economy Moving from theory to practice/The circular economy Moving from theory to practice.ashx.

Nguyen, T.P. (2016) Pesticide use in agricultural production in Thailand, food and fertilizer technology center agricultural policy platform. [Online] Available at: http://ap.fftc.agnet.org/ap_db.php?id=727 [Accessed 10 March 2018].

Office of the Royal Development Projects Board (ORDPB) (2004) The royal development study centres and the philosophy of sufficiency economy. Paper presented at the Ministerial Conference on Alternative Development: Sufficiency Economy. Bangkok.

Rigg, J., Salamanca, A. and Parnwell, M. (2012) Joining the dots of agrarian change in Asia: a 25 year view from Thailand. *World Development*, 40 (7), 1469–1481.

Ryratana, S. (2000) Sufficiency economy. The 1999 TDRI Year-end Conference. Thailand Development Research Institute. [Online] Available at: https://doi.org/10.1355/sj26-lc

Sangsuriyajan, T. (2011) Sufficiency economy as the model of Thailand's community development. *International Journal of Humanities and Social Science*, 1 (5), 74–82. [Online] Available at: www.ijhssnet.com

Sartori, S. (ed.) (2016) *Advancing the Circular Economy in Asia*. SWITCH-Asia Network Facility. [Online] Available at: http://www.switch-asia.eu/fileadmin/user_upload/SCREEN_final_singlepages02.pdf

Singsuriya, P. (2017) Sufficiency economy and backdated claims of its application: Phooyai (Headman) Wiboon's agroforestry and self-narrative. *Journal of Asian and African Studies*, 52 (6), 798–826.

Sriboonchitta, S. and Wiboonpoongse, A. (2008) Overview of contract farming in Thailand: lessons learned. ADB Institute Discussion Paper No. 112. Tokyo. [Online] Available at: http://www.adbi.org/discussion-paper/ 2008/07/16/2660.contract.farming.thailand/

Stirling, A., Leach, M. and Scoones, I. (2010) *Dynamic Sustainabilities, Technology, Environment, Social Justice, Pathways to Sustainability*. [Online] Available at: http://sro.sussex.ac.uk/15485/

Sufficiency Economy Movement Sub-committee, O. of the N. E. and S. D. B. (2007) *Sufficiency Economy Implications and Applications*. Bangkok: Office of the National Economic and Social Development Board. [Online] Available at: http://doi.org/10.1355/sj26-lc

Toop, T.A., Ward, S., Oldfield, T., Hull, M., Kirby, M. and Theodorou, M. (2017) AgroCycle: developing a circular economy in agriculture. *Energy Procedia*, 123, 76–80. [Online] Available at: http://doi.org/10.1016/j.egypro.2017.07.269

United Nations Development Programme (2007) *Thailand Human Development Report 2007 Sufficiency Economy and Human Development*. Bangkok: United Nations Development Programme.

Vandergeest, P. (2012) Deagrairanization and re-agrarianization: multiple pathways of change on the Sathing Phra Peninsula. In Rigg, J. and Vandergeest, P. (eds.) *Revisiting Rural Places Pathways to Poverty and Prosperity in Southeast Asia*, pp. 135–156. Singapore: Nus Press.

Walker, A. (2008) *Royal Misrepresentation of Rural Livelihoods – New Mandala*. [Online] Available at: www.newmandala.org/royal-misrepresentation-of-rural-livelihoods/ [Accessed 19 April 2017].

Walker, A. (2012) *Thailand's Political Peasants: power in the modern rural economy*. Madison: University of Wisconsin Press.

Watabe, A. (2016) Pro-rural policies and people's capacity to aspire: observations of the use of financial resources in the villages of Thailand. *Journal of Human Security Studies*, 5 (1), 19–39.

Win, H.E. (2017) Organic agriculture in Thailand, food and fertilizer technology center agricultural policy platform. [Online] Available at: http://ap.fftc.agnet.org/ap_db.php?id=734 [Accessed 10 March 2018].

Wittayapak, C. (2012) Who are the farmers? Livelihood trajectories in a northern Thai village. In Rigg, J. and Vandergeest, P. (eds.) *Revisiting Rural Places Pathways to Poverty and Prosperity in Southeast Asia*, pp. 211–228. Singapore: Nus Press.

PART V

Conclusion and outlook: circular economy approaches for the Sustainable Development Goals

12
CONCLUSION

Pathways to an inclusive circular economy

Patrick Schröder, Manisha Anantharaman, Kartika Anggraeni and Timothy J. Foxon

The contributions of this book have highlighted the potential of the circular economy (CE) to bring about more sustainable lifestyles and green industrial development in the Global South, while also making explicit that the CE is contested terrain with complex politics. If the CE is to be a generator of inclusive well-being that addresses not only environmental concerns, then an explicit consideration of equity and inclusion is necessary. A failure to consider power and authority in the CE might mean that it generates opportunities solely for big players in industrial development; and it would fail the poor, as has the previous linear development paradigm.

The contributions of this book also reveal that some of the core arguments and issues of the politics of green transformations – multiple pathways, different narratives of transformations, institutional contexts and political alliances – also pertain to the circular economy. Specifically, the chapters of the book demonstrate that:

1. There are multiple narratives and versions of the circular economy; it means different things to different people. There are different circuits and scales of the circular economy where diverse sets of actors and formal and informal institutional arrangements intersect.
2. New alliances consisting of civil society organisations, forward-looking entrepreneurs, media organisations, labour unions, progressive governments and other groups are emerging, pushing for a range of different circular solutions for a variety of interlinked developmental and environmental challenges.
3. The dominant discourse on the circular economy has largely focused on material cycles, resource security and big business opportunities with issues of power, politics and legitimacy receiving limited analytical attention.

4. Without specific consideration of the social implications there will be winners and losers in the circular economy, depending on whose version becomes the dominant and prevailing policy paradigms. The losers will not only be those who fail to make the transition from the linear to circular system, but potentially also marginalised groups already working in the informal CE sectors. They potentially will be pushed out as the business case for circular practices becomes stronger and attracts bigger players with capital and political backing.

In this conclusion section, we first synthesise the chapter contributions to discuss the potential of the CE to accelerate sustainability transitions via their contributions to realising the UN's Sustainable Development Goals. Clearly the concept and practice of CE has much to offer, but what it offers and to whom will in many cases be a question of politics, as we explore in this conclusion. We end by identifying some potential pathways to an inclusive circular economy, emphasising the importance of representation of diverse constituencies in the CE and cautioning against techno-utopianism without adequate consideration of social dynamics.

Potential: circular economy contributions to the UN Sustainable Development Goals

Mainstream discussions of the circular economy, while including discussions of economic growth, materials recycling, and employment opportunities, have thus far lacked explicit reference to the UN Sustainable Development Goals (SDGs). The UN SDGs were adopted in 2015 by the UN member countries and are emblematic of the UN's 2030 Agenda on sustainable development. Although the CE is not mentioned explicitly in the SDGs, circular economy practices could assist in the implementation of the SDGs, as some scholars and practitioners have argued. For example, Gower and Schröder (2016) identified some of the direct and indirect contributions CE makes to SDGs through materials recycling, industrial symbiosis, remanufacturing and repair, and anaerobic digestion and biogas production. They argue that CE approaches can reduce poverty by protecting economic growth from resource price increases and volatility as well as reducing the tension between goals relating to growth and pollution.

A conceptual analysis of potential inter-linkages between SDG targets and CE practices by Schröder, Anggraeni and Weber (2018) showed that strong relationships could exist between CE practices and the targets of SDG 6 (Clean Water and Sanitation), SDG 7 (Affordable and Clean Energy), SDG 8 (Decent Work and Economic Growth), SDG 12 (Responsible Consumption and Production) and SDG 15 (Life on Land). However, there will inevitably be trade-offs between CE waste management approaches and SDG targets, particularly the ones with a focus on decent work, safe working environments and human health, if no additional measures to ensure positive social outcomes are implemented.

Furthermore, Preston and Lehne (2017) highlight that CE practices can support the SDGs through extending product/material life cycles, changing utilisation patterns, looping through additional use cycles and using renewable, recyclable or biodegradable materials. They also highlight the current window of opportunity to align the CE agenda with the implementation of the SDGs. So far, however, only few detailed and integrative strategies aligning the CE and the SDGs have emerged from governments and international policy-making organisations.

Several contributions in this book show the many possible synergies between the circular economy and the SDGs which can contribute to creating such integrated strategies. We start with SDG 12 'Sustainable Consumption and Production', which can be regarded as the core goal for the circular economy and sustainable lifestyles. The approach of reduce, reuse, recycle (3Rs) promoted through Target 12.5 has the most direct link to the CE – without the CE it will be impossible to achieve this 2030 waste reduction target. As the various contributions of this book clearly show (Chaturvedi, Gaurav and Gupta, this book; Noble Gonzales, this book), the informal sector and waste pickers operating in cities worldwide already make a significant contribution towards this target.

Target 12.8 concerns sustainable lifestyles in harmony with nature, an important aspect to consider as the role of consumers in the CE is redefined. In the particular case of marine plastics pollution, tackling unsustainable lifestyle issues through increasing awareness and providing information are hugely important to promote sustainable lifestyles and can shift political debates, as we have shown in Chapter 3 (Schröder and Chillcott, this book). The example of Thailand, as discussed in Chapter 11 (Watabe, this book), shows that lifestyle-related issues are closely linked to food choices and organic farming practices, and wider discussions about the viability of alternative economic paradigms such as the Thai Sufficiency Economy.

Moving to measures and strategies to implement SDG 12 and SDG 8 through CE practices could take form via Target 12.1 and Target 8.4 through the implementation of the 10-Year Framework of Programmes (10YFP) on Sustainable Consumption and Production (SCP). The efforts which are being made to realise the two targets in turn depend on the progress of the 10YFP and how 'decoupling' is viewed by the 10FYP in relation to SCP.

The opportunities for economic development through CE practices also link SDG 8 and SDG 1 ('Eradicating Poverty'), especially the initial Target 1.1 to halve the number of those living in poverty by 2030. Economic development through CE contributes to poverty alleviation, albeit in an indirect way. Other targets to implement 'nationally appropriate social protection systems and measures for all' (Target 1.2), to 'build resilience of the poor' (Target 1.4) are not addressed by CE practices. Additional efforts are required to make CE solutions more inclusive and consider the role sustainable livelihoods play in poverty eradication programmes and policies, and the kind of decent circular job creation strategies assumed in SDG 8.

With regard to the wider scope of SDGs, the concepts of circularity and SCP are key to both implementation and integration of the SDGs, as the goals, sectors and dynamic forces involved are interdependent and cannot be effectively reached as separate silos of competing activities. 'It's difficult to see how we could reach the Sustainable Development Goals with our current economic model', says European

Commissioner for Environment Karmenu Vella (2016), noting that 'we can change' but that 'this requires moving away from our linear economic model to a more circular pattern, where waste becomes a thing of the past'. Most of the SDGs would benefit from the CE and SCP approaches.

Another obvious goal for the CE is SDG 9 'Industry, Innovation and Infrastructure', noting in particular targets to develop quality, reliable, sustainable and resilient infrastructure (Target 9.1) and to upgrade infrastructure and retrofit industries to make them sustainable, with increased resource-use efficiency and greater adoption of clean and environmentally sound technologies and industrial processes (Target 9.4). The example on industrial symbiosis in China's industrial parks (Chen, Song and Anggraeni, this book) shows how this can work in practice. However, the CE is not mentioned among the targets given here, leaving questions as to how much linear economy thinking and habits still shape policymakers' definitions of 'clean', 'sustainable', and 'resilient'.

Less obvious is the link between SDG 5 'Gender Equality' and the CE. Whilst in the mainstream discourse on CE the issue of gender equality is a non-topic, the book contributions showed that CE can become an opportunity for empowerment and social inclusion for women, as in the case of upcycling initiatives in Indonesia (Bebasari, this book). The issue of disposable menstrual products shows that gender issues and cultural issues of concealment around menstruation need to be addressed to make the CE work, an issue mainly unexplored in the mainstream CE debates and highlighted through the case of the Argentinian zero waste framework (Gaybor and Chavez, this book).

Politics of the circular economy

'Questions surrounding what counts as green, what is to be transformed, who is to do the transforming, and whether transformation, as opposed to more incremental change, is required are all deeply political' (Scoones, Leach and Newell, 2015). A growing recognition of the political nature of sustainability and green transformations brings us to the second major emphasis of this conclusion: the circular economy is political. What counts as circular and who gets to decide? Who gets to participate in the circular economy and in what ways? These are key questions that need to be considered as the circular economy moves out of the shadows and onto the centre stage in global conversations on sustainable development.

A key purpose of this edited volume was to document the diversity of practices and initiatives that reflect principles of circularity in resource flows and material cycles, and through this to demonstrate that there is no one version of the CE. Rather, it means different things to different people. There are different narratives about what the circular economy is and how to bring it about. These varied pathways to the circular economy have emerged in different spaces, with some of the most-cited conceptualisations forwarded by think tanks, governments and academics in the Global North. At the same time, alternative visions of and pathways to the circular economy are visible in parts of the Global South, as documented in the contributions to this book.

This book shows how CE practices are being realised in varied domains, ranging from traditional agricultural practices of small-scale agriculture in India compared to industrial symbiosis in China's industrial parks. One key insight from several chapters is that some CE practices pre-date the official proclamations of 'The Circular Economy' by thought-leaders like the Ellen MacArthur Foundation. These practices in many cases are attributable to indigenous epistemologies (Sharma and Joshi, this book) and the resource-conservation ethic of the urban poor as described in Chapters 2, 4 and 5. Reuse, waste recovery, upcycling and resource recycling practices are ubiquitous and longstanding in the cities of the Global South, but the CE as a concept is a relatively new entrant. These chapters highlight the ways in which the theory of CE can be used to systematise and scale these vernacular circular practices to increase environmental and social impacts. Similarly, the case study in rural India demonstrates that traditional practices that have been crowded out by resource-intense and industrialised agriculture can be revitalised using CE ideas, which lend them a veneer of being modern.

At the same time, several chapters demonstrate that we cannot celebrate traditional practices in the Global South without critically examining them from a resource and justice perspective. For instance, the example from Thailand (Watabe, this book) shows that conventional agricultural practices are in many cases not circular, and there can be tensions between new circular livelihoods based on organic farming practices and livelihoods based on conventional agricultural practices. This case and others also demonstrate that the distinctions between what is circular and linear, what is traditional and modern, what is rural and what is urban etc. is contested and in flux. These examples point to a necessity to document more case studies of circular practices in the Global South in a manner that pays attention to who are the holders of the knowledge, who are the practitioners, what is being practiced and what the consequences are for reducing material and energy throughput and generating inclusive well-being.

Diversity in ideas about what CE is or how it is practiced is not in itself a bad thing, as has been more generally documented in the green transformations literature. However, there is always the danger that some conceptualisations of 'green' or 'circular' become dominant and crowd out other ways of thinking and doing. This poses two issues: first, not all ideas work in all contexts, so if some conceptualisations and practices become the norm, we might lose the knowledge generated from grassroots innovations and experimentation. These grassroots initiatives are often more applicable in local contexts, in the Global South as well as the Global North, and align more closely with sustainable lifestyle initiatives. Furthermore, dominant groups in society tend to forward conceptualisations that become dominant, and these often favour already powerful actors. Therefore, depending on which narratives and pathways become dominant, there will be a different set of winners and losers.

Questions about winners and losers are important. Not everybody will be a winner in the transition to the circular economy. So should a CE help those who are already winning win more, or create new opportunities for those who have lost out in the transnational techno-capitalist economy? What will this mean for communities and their livelihoods in the Global South? One example where these tensions are very apparent is around the discussion of whether waste pickers should have preferential access to new CE opportunities in waste management in the cities of the Global South. Should the purpose of advancing the CE also be focused on improving waste picker livelihood conditions or should it be more 'neutral', allowing new corporate actors to move into the recycling business and capture these new economic opportunities? To avoid these trade-offs, which CE approaches can create win-win outcomes with benefits for all?

In addition to asking questions about the knowledge politics of the circular economy, we also have to consider the question of who will be the advocates of the CE. How realistic is it to expect that large corporate actors, having all the means and capacity to shift to CE and who are benefitting from the current linear model, will become the champions of the circular economy? For example, can we trust big consumer-facing brands like Coca Cola, PepsiCo, Nestlé and others who are to a large degree responsible for the current marine plastic pollution crisis to provide the solutions? If not, then who are the responsible stewards of the CE? More importantly, who should be the arbiter of political decisions around the CE? Is this the role of the government? Alternatively, should 'free-market' principles preside?

Finally, the deeply political question of whether a circular economy represents growth-as-usual or should be seen as opening up alternate trajectories of societal progress remains open. As noted in the Introduction, this issue is closely related to the question of the value of natural resources and social benefits. As many of the chapters argue, the wider resource and social benefits of many CE solutions are not captured within current economic frameworks, so that a narrow focus on economic growth could potentially exclude these solutions. This suggests that greater attention to natural and material resource flows within the CE needs to be matched by similar attention to wider measures of social wellbeing (Raworth, 2017; Lamb and Steinberger, 2017).

Pathways towards an inclusive circular economy: the role of different actors in green transformations

The previous section demonstrates that the CE is contested terrain composed of a variety of actors who have different priorities, stakes and needs. The green transformations framework helps us think through the roles and interests of these different actors. Scoones (2016) identifies four distinct pathways of transformations that are driven by different processes, namely state-led, technology-led, market-led, and citizen-led processes. These four pathways are also relevant for the transformation from a linear to a circular system, as has been demonstrated by the various contributions of this book.

State-led transformations are a key pathway for realising an inclusive circular economy, as the successful transformation from a linear to a circular system will not only require instigation by governments, but also have to be politically sustained for long periods of several decades. Furthermore, policies promoting CE will encounter opposition from other actors such as business with stakes in the perpetuation of the current linear model. An example is the issue of policy solutions to tackle single-use plastics, as was discussed in Chapter 3 (Schröder and Chillcott, this book). The EU with the Circular Economy Package and China with the Circular Economy Law are leading the way politically. Nevertheless, other governments are becoming proactive as shown in the contributions of this book. The case study of Colombia, discussed in detail in Chapter 7 (Garcia and Cayzer, this book), is a case in point: while Colombia still lacks a coherent strategy to guide the transition to a CE, the government is designing new policies to shift away from the current economic model that strongly relies on natural resources extraction. Similarly, African governments are beginning to implement a range of new policy measures and have initiated the African Circular Economy Alliance, with the aims of tackling environmental challenges and impacts on human health arising from e-waste through extended producer responsibility policies (Desmond and Asamba, this book).

Technology-led transformations have already been identified as an important element of the CE. As the experience of the Tianjin industrial park shows (Chen, Song and Anggraeni, this book), for the transformation of existing large industrial clusters into more circular systems based on industrial symbiosis, big data and information technologies play an increasingly important role. Research on the links between Industry 4.0, digital manufacturing and the CE is only in its infancy (Jabbour et al., 2018), but the potential implications for developing countries' manufacturing and employment will be significant. Although this topic has not been explored in this book, the close link between Industry 4.0 and the CE could lead to negative social outcomes, e.g. job losses in labour-intensive manufacturing sectors. We also showed that products, technologies and CE practices have a gender dimension, and social taboos around issues such as menstruation pose obstacles to finding sustainable solutions for single-use menstrual management technologies (Gaybor and Chavez, this book). Finding solutions to the complex problem of marine plastics pollution will unquestionably require new technological innovations for packaging materials, but to close plastics leakage will, in addition, require deep changes in social practices of plastics usage (Schröder and Chillcott, this book).

To date, the CE has been dominated by *market-led transformations* in which private sector actors led the transformation, especially big businesses. They have recognised the enormous business opportunities the CE holds. However, a business-driven transformation is likely to have several blind spots, in particular concerning issues of social inequality or workers' rights protection. Will large corporate actors, who are benefitting from the current linear model, become the champions of the circular economy without incentives or sanctions? Indeed, as several chapters on this book demonstrate, it is more likely that CE solutions will come from the numerous new start-ups, innovators and SMEs (Desmond and Asamba, this book) trying to address the pressing economic and environmental issues. However, the research about the municipal waste management and recycling sectors in India shows that a market-led transformation is in many cases not as effective as organised waste picker cooperatives. There is also evidence that current recycling markets on the local, national and global levels are hugely flawed and will require serious interventions to increase the circularity of materials and close leakage points. In China, there are indications that local recycling markets negatively impact on attempts to set up and maintain industrial symbiosis networks in industrial parks (Chen, Song and Anggraeni, this book), limiting the potentials of circularity on industrial level.

Society-led transformations: Many contributions of this book pay special attention to society-led circular economy pathways. Alliances of civil society, church organisations and informal sector workers have been important in the creation of cooperatives of waste pickers in Brazil and India (Noble Gonzales, this book). The transformations to circular systems in rural areas seem to be largely society-led, as the examples from rural communities in Rajasthan and Thailand show. The challenge for small-scale society-led transformations is that unless they create some form of positive feedback through their actions or ideas to scale-up and replicate successful models, they will not be able to lead to significant transformations. The example of society and media-led action on plastics pollution in the UK context shows that social initiatives and campaigns can be powerful and lead to political action and commitments by governments.

The road ahead: making the CE work for human development

The purpose of this edited volume is to document the diversity of practices and initiatives that reflect principles of circularity in resource flows and material cycles, and through this to demonstrate that there is no one version of the circular economy. The implementation of the Sustainable Development Goals (SDGs) offers an opportunity to move beyond small and self-interested political agendas and advance the development of an inclusive CE.

One step towards this is objective is to establish closer connections and cooperation models between the Global South and the Global North. The CE is not new to the Global South and many circular innovations are emerging which can also be used as reference point for actors in the Global North. As the CE matures as a field of practice and inquiry, the next stage of both theorising and practice should focus more explicitly on the social impacts of CE practices in respect to reducing inequalities and increasing inclusiveness.

It is necessary to develop a more sophisticated and holistic concept of the CE which will include social-economic and political elements of the transformation from linear to circular economic models, combined with human development aspects from the social sciences and development studies. To complement the technological-material focused model that is primarily based on principles of industrial ecology and resource efficiency, this book brings in explicit links with the human dimensions.

With this book we hope to have contributed to increasing knowledge and understanding among the CE research community of the missing social and human dimensions in current CE discourse, and second, to familiarise the international development community with the multiple approaches of CE that are emerging. This will advance the options for adopting CE practices and innovations in international development programmes and for the process of implementing the SDGs.

References

Gower, R. and Schröder, P. (2016) *Virtuous Circle: how the circular economy can create jobs and save lives in low and middle-income countries*. London and Brighton: Tearfund and Institute of Development Studies.

Jabbour, A., Jabbour, C., Filho, M. and Roubaud, D. (2018) Industry 4.0 and the circular economy: a proposed research agenda and original roadmap for sustainable operations. *Annals of Operations Research*, 270, (1–2), 273–286.

Lamb, W. and Steinberger, J. (2017) Human well-being and climate change mitigation. *WIREs Climate Change*, 8, e845.

Preston, F. and Lehne, J. (2018) *A Wider Circle? The circular economy in developing countries*. Briefing December 2017. London: Chatham House.

Raworth, K. (2017) *Doughnut Economics: seven ways to think like a 21st-century economist*. London: Random House.

Schröder, P., Anggraeni, K. and Weber, U. (2018) The relevance of circular economy practices to the sustainable development goals. *Journal of Industrial Ecology*. [Online] Available at: https://doi.org/10.1111/jiec.12732

Scoones, I. (2016) The politics of sustainability and development. *Annual Review of Environment and Resources*, 41, 293–319.

Scoones, I., Leach, M. and Newell, P. (2015) *The Politics of Green Transformations*. London: Routledge.

INDEX

Page numbers in *italics* denote an illustration, **bold** a table, n an endnote

Adenso-Díaz, B. 97
Adler, Alfred 13
Africa: CE initiatives 10–11, 153, 156, 168–169, 209; CE transition gains and constraints 153–157, **158**; Ghana's repair businesses 153, 157; imported textile tariffs 12; Kenya, CE transition progress 157–164; Nigeria, CE initiatives 156, 169; Rwanda, CE initiatives 156, 168, 169; solar power market 163; South African mobile phones and CE 164, **165–166**, 166–167, *168*
African Circular Economy Alliance (ACEA) 11, 156, 169, 209
African Circular Economy Network (ACEN) 157, 169
Argentina: Buenos Aires's zero waste framework challenges 99–107; sanitary waste disposal 97

Basel Convention 44
Bharadwaj, S. 98, 107
Bhumibol Adulyadej, King 189
biomimicry 4
bioplastics, seaweed based 51–52
Blue Planet II, policy influencer 45–47, *45–46*
Bocken business model application 76, **77**, **80–81**, 84
Boulding, Kenneth 4
Braungart, Michael 4

British Plastics Federation 48
Brody, Steven 100
Brundtland Commission 1987 12

Cecere, Grazia 96–97
Chambers, Robert 13
Chatham House 10, 204–205
Chaturvedi, Ashish 38–39
Chertow, Marian 135
China: CE Acceleration Policy 134–135; Circular Economy Promotion Law 114–115, 135, 209; industrial parks (IPs) and CE potential 135–136, 147–149, 209; League of National Economic-Technological Development Zones (NETDZs) 145, 147; recycling industry, barrier to CE 141, 148, 210; TEDA Eco Centre, operations and networks 136–137, *138–140*, 140–145, **141**, *142–143*, *147*; waste import ban 12, 50, 52; Whole Process Management of General Industrial Solid Waste project 141–143, *144*, *146*
Chintan Environment Research and Action Group 30, 61–62
circular economy and African transitions: development, positive actions 155; economic value inequalities 153–154; green economy developments 153; initiatives, case studies 157, **159**; Kenya, green economy and beyond 157–158,

160–164; policies and regulations 157, **158**; power relations and inequality 154–155; South African mobile phones and CE 164, **165–166**, 166–167, *168*; stakeholder networks 156–157

circular economy (CE): business model (Bocken) **77**; China's adoption 134–135; concept, roots to modern 4–5; corporate framing as restrictive 26; development ideology and power relations 11–12, 203–204, 206–208; economic value, interpretation issues 7–8, 115, 153–154, 156, 207–208; Global North adoptions 5, 114–115, 152; green transformations perspective 6–7, 58, 76–77, 208–210; inclusive development 15–16, 210–211; industrial symbiosis approach 135; pathways of transformations 208–210; principle objectives 3, 152; resource productivity and rebound effect 8–10; social and institutional issues 9; Sufficiency Economy comparisons 190; Sustainability Development Goals (SDGs) 77, 204–206; transition considerations 5–6; UK government discourse 49–50

Circular Economy Club 156

circular economy, transition readiness, Colombia: business input and SME support 128–129, 130; CE education absent 125–126; digital infrastructure challenges 125, 130; enabling framework components 118–121, *119*, 129; financial barriers 127–128; methodology 118; non-renewable dependency, limiting factor 121, 124–125, 130; readiness evaluations 121, **122–123**, 130; recycling, safety and finance issues 127, 130

civil society organisations (CSOs): Indonesian waste campaigns 75–76; VAAGDHARA, India 175–176

Coca Cola 48

Colombia: biocapacity decreases 113, 117; CE as development strategy 10, 209; CE transition readiness study 118, 121–130, **122–123**; economy, current state 115–116, **116**; Green Growth Mission 118, 124; National Development Plan for 2014–2018 118, 124; National Policy for Sustainable Consumption and Production 117–118, 124, 125; sustainable development, disjointed actions 117–118, 121, 124–125; urbanisation and degradation issues 117

Conway, Gordon 13

Cornwall, Andrea 87

corporate social responsibility (CSR) 47–48
Côté, Raymond 135
cradle to cradle design 4, 96

Davidova, Sophia 185
Davidson, Gary 97
disposable menstrual management technologies (DMMT): Buenos Aires waste issues 100–101, 103, 106; lifetime use estimates 100; waste disposal studies 97–98
Dreamdelion **80**, 81, 82–86, 89n3
Drowning in Plastic (BBC) 50–51

Ellen MacArthur Foundation: 'Butterfly' diagram 155; circular economy and GDP 8; circular economy concept 4, 178–179; technical/material development focus 7, 26, 32, 49, 120
endogenous rural communities, India: linear agriculture models, negative impacts 176–177, *177*; malnutrition and marginalisation 175–176; nutrition-sensitive farming system (NSFS) study 177–186, *178*, *181–182*
Esposito, Mark 49–50
European Union (EU): Action Plan for the Circular Economy 5, 7–8, 49, 154; Circular Economy Package 58, 114, 167, 209; e-waste recycling 12; extended producer responsibility (EPR) 52, 155, 157; resource efficiency and security agenda 32, 114, 154; Resource Efficient Europe 114; SWITCH-Asia Programme 137, 190
e-waste recycling: EU policy 12; Indian urban sector 38–39; Kenyan initiatives and policy 162; mobile phones, CE solutions 166–167, *168*; mobile phones, South African issues 164, **165–166**
extended producer responsibility (EPR): African economies 157; Colombian programs 121, 125, 127; EU policy 52, 155

Fairphone 155, 167
Felitti, Karina 104

Gaventa, John 28–29
Geissdoerfer, Martin 5
gender equality: CE initiative potential 15–16; reusable menstrual management technologies, access to 101–102, 103–106, 107; upcycling, women's initiatives, Indonesia 79, **80–81**, 81–89; women's empowerment 77–78

Gerba, C.P. 98
Giddens, Anthony 13
Global Alliance for Incinerator Alternatives (GAIA) 30, 32–33
Global Alliance of Waste Pickers 38
Global Footprint Network 113
global value chains (GVCs) 154
González-Torre, Pilar 97
Gower, Richard 168, 204
green economy: African developments 153; Kenya's strategy 157–158, 160; UK government actions 47; Western definition 14
green industrial development: CE transition readiness, Colombia 121, **122–123**, 124–130; Global South opportunities 10–11; industrial parks (IPs), China 135, 149; *see also* industrial symbiosis, TEDA Eco Centre, China
green transformations: actors, vertical linkage 29, 68; circular economy in context 6–7, 58, *58*, 76–77, 208–210; stakeholder alliances and priorities 68–71; upcycling initiatives, Indonesia 85–86
Guerrero, Lilliana 97

Houppert, Karen 107
Howard, Courtney 100

India: informal waste pickers, Delhi 60, *61*, 62–63, *66*, 68–70; nutrition-sensitive farming system (NSFS) study 177–186, *178*, *181–182*; single-use plastics ban 51; urban resource flows and degradation 25; waste management, CE transition narratives 27–30, **29**, 32–39, **37**
Indonesia: plastic waste, policy and activism 75–76, 79; upcycling, women's initiatives 79, **80–81**, 81–89
industrial symbiosis, TEDA Eco Centre, China: CE promotion and online services 143–145, *147–148*; data management and facilitation 137, 140–141, 140, **141**, *142–143*; founding role 136; industrial parks (IPs) strategies 135–136, 147–149, 209; operations and networks 136–137, *138–139*; waste management, labelling and evaluation 141–143, *144*, *146*
Industry 4.0 and CE 209
informal waste pickers: activity estimates 57; CE initiatives and inclusion 15–16, 38, 67–68, 70–71, 154–155, 210; dismissive criticisms 59–60; effective recovery systems 58–59, **59**, 210; Indian city conflicts 34–35

informal waste pickers, Delhi, India: activity rates and workforce 60, *61*; finances and privatisation impacts 62–63, 68–69; improved conditions, NGOs role 60–62; inclusion process failures 69–70; waste management, closing loops *66*
informal waste pickers, São Paulo, Brazil: finances and infrastructure 64–65, 69; municipality integration 63; operations and activism 63–64, 67–68; stakeholder alliances and laws 65–66, 70; waste management, closing loops *66*, 66–67

Jackson, Tim 96

Kabeer, Naila 78
Kenya, CE transition progress: e-waste recycling and regulation 161–162; green economy strategy 157–158, 160; plastic bag ban 51; rural electrification and solar power 162, *163–164*; urban waste, dumping to recycling 160–161
Kirchherr, Julian 152
Klein, J. 160

Leach, Melissa 28–29, 68
Lehne, Johanna 10, 204–205
Lemille, Alex 155
Lieder, Michael 119
lifestyles, term origin 13
Liu, Changhao 135
Lucas, Caroline 49

MacArthur, Ellen 49
marine plastics pollution: corporate social responsibility 47–48; Indonesian discharges 75; international convention limitations 44; plastics industry's agenda 47–49; reduction initiatives, Global South 51–52; research gaps 43–44; transformative alliances, role of 52–53, 210; UK policy action, "Blue Planet Effect" 45–47, *45–46*; waste management inadequacies 50–51; waste mismanagement and policy failures 43–45
material cycles, closing of: actor-based, multi-scaled framework 26–29, 30, *31*, 39–40; city government challenges 34–35; economic policy barriers 32–34; intra- and inter-level alliances 35–36, **37**, 40; power dynamics 37–39; STEPS pathway analysis 28–29
McDonough, William 4
menstrual management technologies: culture of concealment 103–104, 107;

disposables, Buenos Aires waste issues 100–101, 103, 106; disposable usage 100; reusables uptake, challenges and potential 101–102, 104–106, 209

Movimento Nacional dos Catadores de Materiais Recicláveis (MNCR) 64

Najam, Adil 11
National Environment Management Authority (NEMA), Kenya 162
Nationally Appropriate Mitigation Action (NAMA), Kenya 161
National Waste and Citizenship Forum, Brazil 65–66
Netherlands Environmental Assessment Agency 5
New Delhi Municipal Council 60, 61–62, 68–69
Newell, Peter 68
Nigeria, CE initiatives 156, 169
Noble, Patricia 154
non-governmental organisations (NGOs): Indian waste management **29**, 30, 60–62, 69–70; informal waste pickers, São Paulo 66, 70; women's CE initiatives, Indonesia 81, 82–83, 84–88
nutrition-sensitive farming system (NSFS) study, India: circular economy, appliance and principles 178–180, 210; design goals 177; farmers' CE approach *178*; farming and resource reviews 182–183; food security and CE 185, 186; national policy barriers 185; participatory learning action (PLA) tools and framework 180–181, *181–182*, 185; positive evaluations 184; practices redesigned/readopted 183–184; sustainable transformations 184–185

Obama, Auma 14
OECD, Colombia evaluated 116–118, **116**, 124–125
organic farming, Thai integration study: adoption group activities, members and opinions 192–195, 197–198; chemical users, profitability as key 196–197, 198; employment diversification 191–192, 195; household use and livelihood maintenance 195–196, 198; impact on Sufficiency Economy 189–190; sustainability in context 190–191, 198

Pakistan, community recycling 51
Patkar, A. 98, 107
Pearce, David 4
PEKKA **81**, 82–88, 89n5

performance/sharing economy 4
Potting, José 5–6
power-cube framework 28–29, 37–39
Pradhan, J. 98
Preston, Felix 10, 204–205

Rashid, Amir 119
Reddy, Rajyashree 15
ReSOLVE framework application 76, **80–81**, 84
Rwanda, CE initiatives 156, 168, 169

São Paulo, Brazil: informal waste pickers 63–68, *66*; sanitary waste issues 97
Sardenberg, Cecília 78, 87
Schröder, Patrick 154, 168, 204
Scoones, Ian 7, 28–29, 68, 208
sharing economy 9
Shihata, Alfred 100
Shi, Lin 135
SiDalang 81–87, **81**, 83–85, 86–87
Simmons, S.L. 98
SMEs (small and medium enterprises): CE transition barriers, Colombia 126, 128–129, 130; industrial symbiosis strategies, China 140, *140*
solar energy market 163–164
Sommer, M. 98
South Africa: mobile phones, waste and CE solutions 164, **165–166**, 166–167, *168*; REDISA corruption case 155
SriFsti 61
Stahel, Walter 120
STEPS pathway approach 28
Stirling, Andy 28–29
Suen, Serena 85
Sufficiency Economy: circular economy comparison 190; organic farming, Thai village responses 192–198; Thailand's development policy 189–191
Sustainability Development Goals (SDGs) 44, 77, 204–206
sustainable lifestyles: concept 3–4, 13–14; diversity challenges 207; nutrition-sensitive farming system (NSFS) study, India 181–185, *181–182*; organic farming, Thai integration study 192–198; Sustainability Development Goals (SDGs) 205; Thailand's Sufficiency Economy philosophy 189–191; upcycling and women's empowerment, Indonesia 84–86
sustainable livelihoods: concept 4, 12–13; nutrition-sensitive farming system (NSFS)

study, India 177–186, *178*, *181–182*; organic farming, Thai integration study 192–196, 197–198; practice adoption challenges 198, 207; Sustainability Development Goals (SDGs) 205
SWITCH-Asia Programme 137

Tearfund 155
Thailand: organic farming, integration study 191–198; rural transitions 188–189; Sufficiency Economy and organic farming 189–191
Thomson, Kenneth 185
Timlett, R. 97
transnational waste: China's ban, effect in UK/Europe 50, 52; plastic exports, shifted problems 50–51; restricted and negotiated policies 11–12
Turner, Kerry 4

UK Plastics Pact 48
UNIDO 10, 137
United Kingdom (UK): 25-Year Environment Plan 45–47, *45–46*, 49; circular economy, political discourse 49–50; National Industrial Symbiosis Programme (NISP) 137
UN legislation/conventions: Conference on Environment and Development 1992 12; MARPOL 44; plastic pollution 52; Rio +20 Summit 44; 'Single-Use Plastics' report 51; UNCLOS 44
upcycle, definition 89n2
upcycling, women's initiatives, Indonesia: capacity building 87–88; CE links, empowerment and green transformation 84–86, 88–89; empowerment activities 83–84; mapping CE initiatives 79, **80–81**, *81*; resource and power dynamics 83; study methodology 79; support and equality issues 86–87; training programmes and marketing 81–83

VAAGDHARA, India: endogenous farmers and food security 175–176; nutrition-sensitive farming system (NSFS) study 177–186, *178*, *181–182*
Vallejo, Maria C. 117
value chains, global: Extended Producer Responsibility legislation 157; material cycles, unequal governance 26, 117, 154; mobile phones, South African issues 164, **165–166**, 166–167, *168*; multinationals and SMEs 120, 153; North/South inequalities 12
Veblen, Thorstein 13
Vyas, Anjali 86

waste management, CE transition: actor-based, multi-scaled framework 26–29, *31*, 39–40; informal sector's significance 27–28, 34–35, 39; local, private and informal sectors 34–35, 38–39; national policy and countercriticisms 33–34; power dynamics 37–39; stakeholder alliances and priorities 28–30, **29**, 35–36, **37**; supra-national perspective 30, 32–33
waste management cycle, closing loops: EU policy changes 5, 49, 58; informal waste picker activity 57, *66*, 66–67, 68
Weber, Max 13
WIEGO (Women in Informal Employment: Globalizing and Organizing) 15, 38
Williams, A.T. 98
Williams, I.D. 97
women's empowerment: circular economy links 77; objectives and process approach 77–78; upcycling initiatives, Indonesia 83–89
World Economic Forum 10–11, 30

Zaman, Atiq Uz 97
zero waste approaches: consumption behaviour studies 96–97; definitions and principles 96; national variations 95
zero waste framework, Buenos Aires: cultural barriers 103–104, 107; disposable menstrual products, waste challenges 100–101, 103, 106; law goals and implementation issues 106; legislative goals and implementation issues 95, 99–100; poor consumer engagement 102–103, 106; reusables uptake, challenges and potential 101–102, 104–106, 107; study methodology 98–99

Printed in Great Britain
by Amazon